A LAND BETWEEN TWO NILES

A LAND BETWEEN TWO NILES

Quaternary geology and biology of the Central Sudan

Edited by
MARTIN A.J.WILLIAMS & D.A.ADAMSON
Macquarie University, Australia

A.A.BALKEMA/ROTTERDAM/1982

To the people of the Sudan

ISBN 90 6191 096 X
Published by A.A.Balkema, P.O.Box 1675, Rotterdam, Netherlands
Distributed in USA & Canada by MBS, 99 Main Street, NH 03079
Printed in the Netherlands

Table of contents

Preface

Three themes run through this book, each of which deserves fuller treatment than we have allowed. The first is the history of the Nile; the second is the degree to which the present Sudanese landscape reflects the operation of former geological processes; the third is the interaction between man and environment, not always to the benefit of either.

In a country where for much of the year water is less plentiful than land, it is natural that great efforts are made to moderate and conserve the flood-waters of the two great rivers which converge at Khartoum. Water control is ineffective without wise use of land, so that a knowledge of what the land can bear is essential to any irrigation project, however large or small. Will the soil become saline? Will the land become waterlogged? Will today's farm be tomorrow's waste land? These are not trivial questions. They are ones that worry Governments and farmers throughout the semi-arid world, as Peter Crabb points out in our opening chapter. In the Sudan, where irrigated cotton from the Gezira provides much of the export revenue, these issues are of national concern. They are also one of the stimuli which gave rise to this book.

A land between two Niles is an interdisciplinary account of the origins and characteristics of the alluvial plains of the lower Blue and White Nile. We have focussed our attention upon this region for several reasons, one of which we alluded to earlier. Although the Gezira plain itself only occupies about one percent of the total area of the Sudan, the high quality long-staple cotton grown there provides nearly two-thirds of the country's total export revenue.

A knowledge of the pattern of surface and subsurface salinity is vital to the present and future development of this region, as indeed is an appreciation of the physical and chemical properties of soils and sediments throughout the central Sudan. These properties, and the soils themselves, are in large measure a depositional legacy of the two Niles, a legacy which we describe in this book.

An understanding of the alluvial history of the Blue and White Nile can tell us a great deal more than why sand dunes or clay swales occupy certain parts of the Gezira today. To unravel the former flow regime of these two great rivers is to begin to understand something of the history of the main Nile downstream, together with its impact upon the prehistory of man.

In fact, situated as it is at the junction of the Blue and White Nile, roughly midway between the Nile delta in Egypt and the mountainous headwaters in Uganda and Ethiopia, the Gezira is a vital link in any attempt to relate the Quaternary alluvial history of Egypt to hydrological fluctuations upstream.

A few lines on the personal background to this book may be helpful. During 1962-64 one of us (M.W.) was engaged in mapping soils along the lower Blue and White Nile with Hunting Technical Services as part of the Roseires Dam Soil Survey project. As part of this work in 1963-64, Ekhlas Abd el Bari of Khartoum University Botany Department made collections of the plants growing along the lower White Nile. A decade earlier, R.H.Gunn had carried out an extensive survey of the soils and plants in the clay plains south of the Gezira on behalf of Sir Alexander Gibb and Partners.

There was a danger that much of this hard-won information about the country would gather dust in Government and company files. It was also obvious that additional fieldwork was necessary if we wanted to know with greater precision when the rivers were high and when they were low. And so, after a quick reconnaissance by the two of us in January 1972, we began work in earnest in the winter of 1972-73, rounding it off with a field-based workshop for botany, geology and archaeology graduates in December 1976. To our joy, we were again helped in this by our friend and colleague Dr Hassan Hag Abdulla from the Gezira Research Station in Wad Medani.

The 1972-73 archaeological excavations at Jebel et Tomat and at Shabona were directed by Professor J.Desmond Clark from the University of California at Berkeley. We have worked closely with Professor Clark in Niger, Ethiopia, India and the Sudan for over ten years. The long, painstaking task of archaeological analysis is now complete, and this work will be the subject of a separate volume on the prehistoric settlement of the central Sudan.

Professor John Vail, who recently completed the enormous task of remapping the geology of the entire Sudan, here describes the geological setting of the Gezira, the stratigraphy of which is discussed by Drs Ismail Elboushi and Yassin Abdel Salam, with whom we examined the outlets of both Niles in December 1972.

Dr Mohamed Obeid Mubarak, now Vice-Chancellor of the new University of the Gezira, and Dr Ekhlas Abd el Bari, Head of the University Herbarium and at all times a source of help and inspiration, have shown how plants capable of withstanding the stresses of their arid environment have been unable to withstand the depredations of man and his herds. The result, outlined by Dr Gerald Wickens for the country as a whole, is a gradual southward retreat of plants and animals.

This book is not a substitute for J.D.Tothill's *Agriculture in the Sudan* (Oxford U.P., 1948), A.J.Whiteman's *The Geology of the Sudan Republic* (Oxford, Clarendon Press, 1971), and Julian Rzóska's *The Nile: Biology of an Ancient River* (The Hague, Junk, 1976). Nor is it simply a more detailed supplement to certain chapters in those three most useful books. It is an attempt to delve below the surface, into the past, using evidence from a variety of disciplines, the better to understand the heartland of the Sudan.

Sydney, December 1980

Martin Williams
Donald Adamson

Acknowledgements

Our debt of gratitude, like the Gezira itself, has its roots in the past. As a novice to the Sudan in 1962-64, one of us (M.W.) recalls with appreciation the unfailing courtesy of the people of the central Sudan, and in particular the hospitality freely proferred by Sayed Idris Habbani, Umda of Hashaba; by the late Sayed Omar Mustafa, Umda of Esh Shawal; and by the late Sayed El Hadi Abdel Rahman El Fadi el Mahdi, who generously made available his private resthouse.

A decade later our paths were to be made smooth by friends old and new. In the University of Khartoum, Dr Ekhlas Abd El Bari and Dr Mohamed Obeid Mubarak of the Botany Department; Professor John Vail, Dr Ismail El Boushi, Dr Yassin Abdel Salam and Dr Salah El Raba'a of the Geology Department; and Dr Asim El Moghraby of the Hydrobiology Research Unit, were at all times towers of strength.

The administrative skills of Dr El Sammani Abdalla Yacoub and Dr Ali Mohayad Bannaga of the National Council for Research helped enormously, bringing us into fruitful contact with Dr Hassan Hag Abdalla of the Gezira Research Station, as well as with our geological colleagues Sayed Abdel Latif Widatalla, Sayed Yusif Sulieman, Sayed Abdulla Hassan Ishag and Sayed El Tayeb M. Saeed, all of the Sudan Geological Survey.

It was a great privilege to work with Professor J.Desmond Clark and his archaeological party in 1973, and we are grateful to Sayed Negm-ed-Din Sherif, Director of Antiquities, for visiting the excavations.

In January 1972 Sayed Awad Medani, Tony and Pat Harris and Professor John Cloudsley-Thompson helped to enlarge M.W.'s knowledge of the Sudan; he thanks them for their help and hospitality.

During January 1976, Frances Williams and the late Dr Bill Morton accompanied us up the Blue Nile, and in December of that year Dr John Gowlett came with us during our survey along the left bank of the White Nile. We thank them for their companionship and help.

Our thanks go also to Mrs Alison Coates, Miss Campbell Donaghy and Miss Ruth Sefton, for typing countless drafts; and to Sharon Knight, Dean Oliver and John Cleasby for skilful drafting of the figures.

Finally, and by no means least, the Khartoum staff of Hunting Technical

Services and of Sir Murdoch Macdonald and Partners were invariably most helpful, for which we thank them warmly, as also Mr V.C.Robertson, Mr Ian Matthews, and Mr Martin Adams. It is a pleasure to record the encouragement received from our own University, and in particular from Professor Alec Mitchell and Professor Edwin C.Webb, the past and present Vice-Chancellors of Macquarie University.

PETER CRABB
Macquarie University, North Ryde, Australia

1

Alluvium and agriculture in the semi-arid world

1.1 INTRODUCTION

The aims of this book are to describe and to interpret the depositional history and Quaternary environments of the Central Sudan, and to demonstrate the significance of such knowledge to the current utilisation of the region and to an understanding of some of its problems. Such problems are by no means unique to this part of the Nile Valley. Thus this first chapter discusses a number of general issues in a wider setting. In the chapters that follow, these issues are discussed in detail with special reference to the alluvial plains of the Central Sudan.

If we define as arid those areas where sustainable agriculture is not possible without irrigation, then some 36 per cent of the total area of the world's continents is arid or semi-arid (Meigs 1953; Dregne 1974; Nir 1974) (Figure 1). Some 13 per cent of the world's population lives in the dry third of the earth, with approximately three-quarters of this number in the semi-arid zone (Amiran 1966). An average population density of about 13 persons per km^2 for this zone tells us little, as the distribution of population and settlement is highly discontinuous and concentrated around the fringes of the zone and in other favoured localities. Some of these may be mineral deposits, as in northwest Australia, Chile and Mauritania, but the real key to a favoured locality is the presence of a permanent (or near permanent) water supply and a potentially irrigable soil. Water, however, is the prime determinant of all economic activity and settlement. Without a reliable water supply, life and activity are dependent on an extremely unreliable rainfall, which can support little more than very extensive nomadic pastoralism and short-period dry farming (Nir 1974). Where a reliable water supply is available, be it from a major river, with a source outside the semi-arid zone, such as the Nile and Indus, or from groundwater, such as the Saharan oases and the large Australian artesian basins, the opportunities for a much wider range of economic activities are greatly increased. It is in such localities that we find the major concentrations of population in the semi-arid zone.

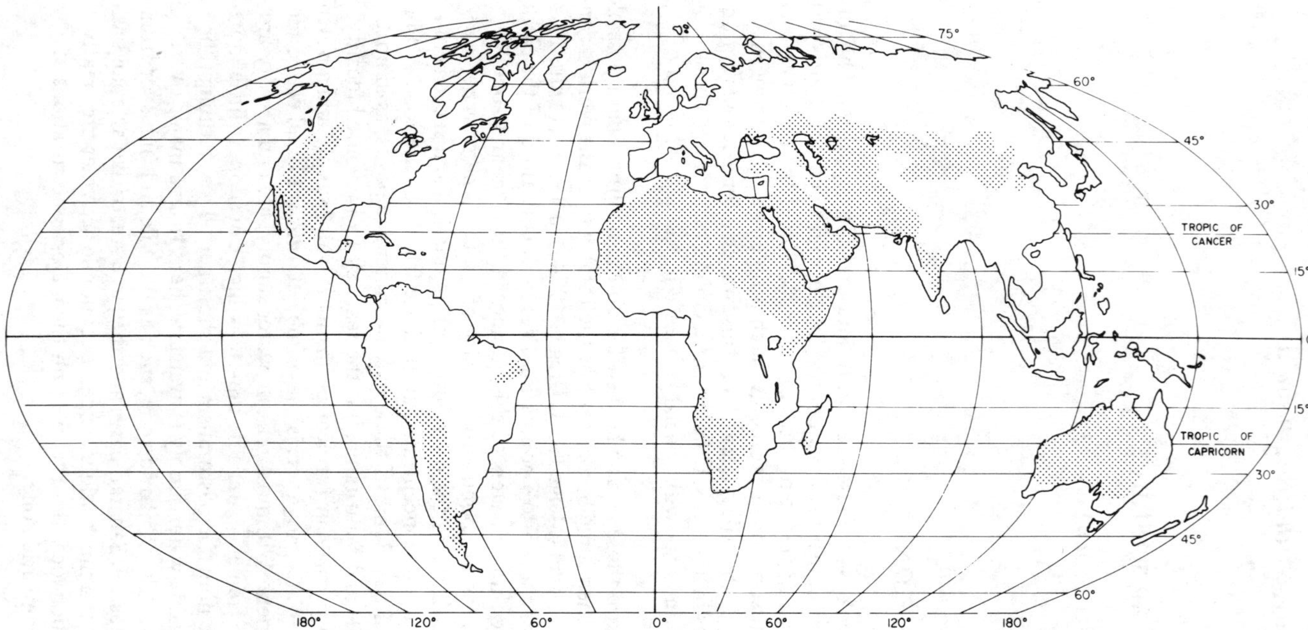

Figure 1.1. The semi-arid and arid regions of the world (adapted from Meigs 1953; Dregne 1974; Nir 1974).

1.2 THE ALLUVIAL PLAINS

Of particular interest and significance are the alluvial deposits of the semi-arid zone, for it is these areas and their agricultural activities that support much of the population resident in the semi-arid zone. These alluvial areas exhibit considerable variation (Cooke & Warren 1973), from the small alluvial fans of the Negev (Evenari 1971), to the vast plains of the Murray-Darling and Diamantina Rivers in Australia. A series of gently sloping alluvial fans, together with their flood-plains, may merge to form one large area of alluvial deposition, as in the Central Sudan and south-eastern Australia (Figures 2, 3). For example, the Riverine Plain of the eastern Murray Basin is considered to be an alluvial plain of medium size, yet it still extends over an area of almost 77 000 km^2 in southern New South Wales and northern Victoria (Butler et al. 1973). Such alluvial plains can provide surface runoff as well as groundwater, though both can be unreliable and problematic in terms of quantity and quality, especially in the smaller areas.

These alluvial plains have been formed by rivers that have their headwaters in higher, wetter areas. For example, both the Australian Riverine Plain and the Gezira Plain in Central Sudan were formed during the Cenozoic Era by rivers flowing from humid uplands into dry lowlands where they built up extensive low-angle alluvial fans consisting of level clay plains, former stream channels, swampy depressions, and sand dunes. Such plains exhibit great variety in terms of their geology, the sediments of which they are formed, and also their geomorphology, in large measure the result of their complex depositional histories and the considerable climatic variations experienced both during and since their formation. The depositional history influences more than the physical characteristics of the soil materials: it also has a bearing on their chemical nature, notably salinity. This is also true of groundwater deposits which show considerable variation in quality. Further, the ability of groundwater to move from one location to another is also largely dependent on the depositional history of the area in question. A lack of water, complex geomorphology, soils and groundwater of varying salinity, a tendency for salinity levels to increase under the prevailing dry climate, through high evaporation as well as through the addition of cyclic salt (Jennings 1955 and 1956), clearly provide a natural environment of some difficulty for human activity.

1.3 IRRIGATION AND AGRICULTURE

But for these alluvial plains, their soils, and especially their water resources, intensive agriculture would be impossible in the arid world. In spite of their inherent limitations, these alluvial plains are areas of high agricultural potential. Using both surface and groundwater resources, areas of alluvial soils have formed the basis of most of the world's large irrigation areas (Jewitt 1966). For thousands of years, irrigation systems of varying degrees of technological complexity and efficiency, have made possible the production of a wide range

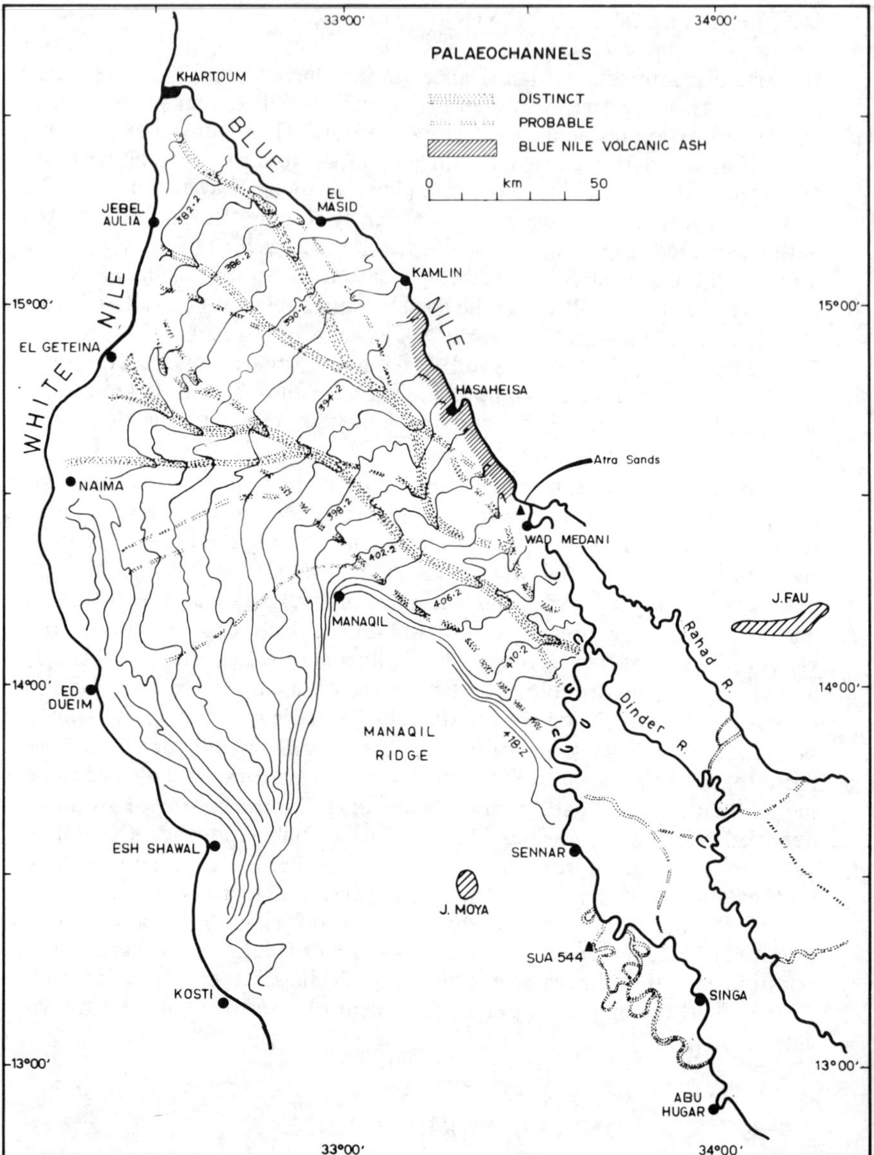

Figure 1.2. The Gezira alluvial plain between the Blue and White Nile rivers, central Sudan (adapted from figures in chapters 7, 9 and 10 of this volume).

of agricultural commodities, notably in the Middle East. As well as providing basic food crops, such as cereals and pasture and fodder for livestock, such irrigation areas are also the source of many high-value commodities, which is not surprising given the added costs of irrigation production. Goods for both domestic and export markets are produced, in some cases of major significance, though perhaps few equal the case of the Gezira's cotton production, which comes from barely one per cent of Sudan's area yet accounts on average for 60 per cent of the nation's export revenue (National Council for Research 1975). The Australian Riverina – so similar to the Gezira in terms of, for example, surface gradient, palaeochannels, clay mantled flood plains, and fixed source-bordering dunes of reworked fluviatile sands that were formed some 15 000 years ago (Figure 3) – accounts for much of the country's irrigated horticultural production and irrigated pasture as well as almost all of its rice crop. In North America, irrigation schemes in the Utah Valley, along the Rio Grande and Brazos River in Texas, and along the Lower Colorado River, are important sources of food and industrial crops (National Academy of Sciences 1968), as are such schemes in parts of northern Mexico (Hernandez 1974). This is also true of alluvial areas on the Peruvian coastal plain, as in the valleys of the Chancay and Rimac Rivers (where rice, sugarcane, vegetables, cotton and fodder are grown), and in the Monte Desert of north west Argentina (where irrigation from the headwaters of the Rio Colorado and Rio Negro supports orchards and vineyards) (Cornejo 1974; Morris 1969). Cereals and cotton are major commodities produced under irrigation in the arid areas of north west China (Vermeer 1977), while irrigated agriculture is also important in the arid areas of southern USSR, as in the valleys of the Syr Darya and Amu Darya, which flow into the Aral Sea (Petrov 1974; Nir 1974). The Nile has long been the life-blood of the Sudan and Egypt, both as a source of water and as the basis of their agriculture. The Tigris and Euphrates have played a comparable role for Iraq and Iran, as has the Indus in Pakistan (Newhouse et al. 1950). The socio-economic significance of such irrigated agriculture in the regional and national economies of the countries concerned cannot be underestimated, especially those that fall largely within the arid world (Downing & Gibson 1974).

1.4 SALINITY

It has already been noted that the alluvial areas of the semi-arid zone suffer from a number of inherent difficulties. Whilst irrigation may be a means of adapting to a particular environment, it also increases the ecological fragility of such areas (Downing & Gibson 1974; Goodall & Perry 1979). As has been observed, irrigation is a difficult art (Batisse 1969). Problems almost inevitably follow irrigation, many of them an accentuation of inherent difficulties noted earlier. One of the first is a rise in the level of the water table, though the use of groundwater for irrigation often results in a lowering of the water table, as in the Indus Valley.

'The rise may rapidly reach a dangerous level even where the original

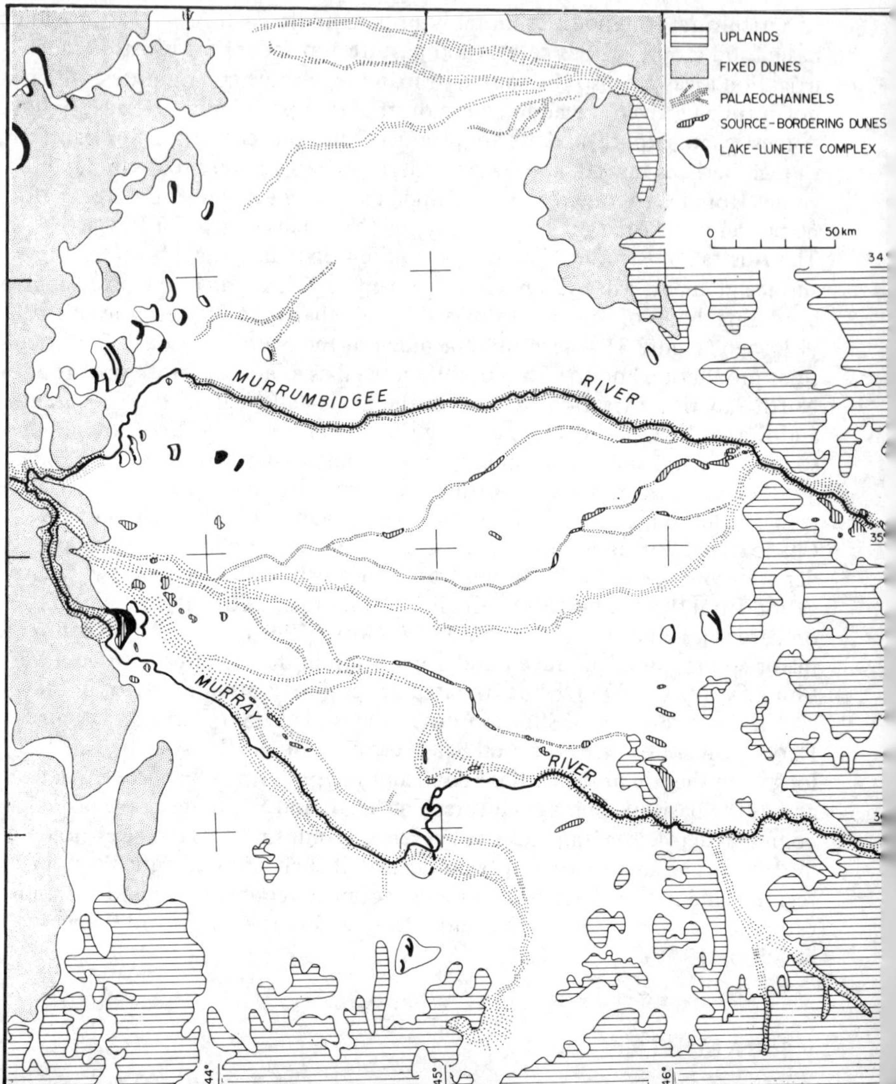

Figure 1.3. The Riverine Plain of the Murray and Murrumbidgee rivers, southeastern Australia (adapted from Butler et al. 1973).

water table was very deep. In the Kerang irrigation district in Victoria, Australia, the groundwater prior to irrigation about 1880 stood 8 metres below the surface: by 1900 it had risen 6 metres, and in 1957 it was everywhere within 1.8 metres of the surface, and under active irrigation higher than this' (Jewitt 1966).

At best, this can result in waterlogging of the soil and even surface water accumulation, to the consequent detriment of plant health and productivity. If, however, the groundwater and/or the soils are saline, the problems are exacerbated, further reducing plant health and crop productivity. In the extreme, high salinity levels in the soil and groundwater can render land barren and completely unproductive, as can be seen in many irrigation areas (e.g. along the River Murray in Australia). Another source of increased salinity can be the irrigation water itself, from both surface and groundwater sources, as salinity levels are increased by such things as evaporation, seepage from canals, and contamination by drainage and other more saline waters. Seepage losses from canals can come to the surface in low lying depressions, where evaporation simply aggravates salinity levels. Evaporation, groundwater seepage, and the addition of drainage water, contribute to a general downstream increase in river salinity (Figure 4), particularly in rivers with considerable irrigation use, such as the Rio Grande, Colorado, and Murray (Gutteridge, Haskins & Davey 1970; Crabb 1969). Some problems can be the result of applying too much water, but where saline water has to be used (and especially on saline soils), crop production is only feasible by applying large quantities of water – in conjunction with an efficient drainage system – to permit leaching and the movement of harmful salts away from the root zones of the plants (Batisse 1976). The problems are many, and they find expression in reduced productivity, lower returns to farmers, and, in many instances, the abandonment of farms. The multiplier effects of such problems clearly have wide socio-economic implications.

The problems are world-wide, and many, if not most, of the semi-arid zone's irrigation systems are experiencing problems of secondary salinization (Worthington 1977). This can be well illustrated in the Riverine Plain and throughout the Australian Murray Valley (Northcote & Skene 1972). Since the latter part of the 19th century, there has been massive investment in irrigation throughout the Valley – over $1 330 million (US) (at early 1970s values) by governments alone (Davidson 1974). Yet problems developed very early, and many schemes were abandoned even before 1910. As well as such matters as the economic size of farms, salinity and waterlogging were problems that appeared early and ones that have continued to exist. They were the result of inadequate research and investigation prior to the establishment of the irrigation schemes (Langford-Smith & Rutherford 1966; Frith & Sawer 1974). Today, salinity is the major problem throughout the Murray system, and it is one that is going to be extremely costly to alleviate, with even greater costs if nothing is done. The most recent study has indicated that over $200 million (US) should be spent over the next 25 years to tackle just the most urgent salinity and drainage problems in the Murray Valley (through saline groundwater interception works, drainage to relieve waterlogging and lower water tables, and improved irrigation efficiency by farmers) (Maunsell & Partners 1979; Anon. 1976). Today, there is no lack of knowledge of the major problems of the Murray Valley's irrigation areas (Gutteridge, Haskins & Davey 1970; McMichael et al. 1976; Maunsell & Partners 1979). It is quite clear that, within the Riverine Plain, for example, salinity and other problems vary considerably from one location to another, and that such variation can be attributed to the detailed depositional history of the area (Butler et al. 1973; Butler 1974). That such problems persist with little remedial action in progress, can in large measure be attributed to the lack of co-ordinated action by the governments of Australia and the three States concerned (Senate Standing Committee 1979). (This is just one consequence of Australia's historical development and the autonomy enjoyed by the states in many areas of jurisdiction, such as agriculture and resource development.) In practice, many of the problems that have been discussed are the result of non-existent or inadequate drainage – to remove saline irrigation water and lower water tables – and the inefficient application of water. As has been observed, 'In reality, quantity, quality and management methods are intrinsically linked' (Batisse 1976). In many instances, more than an efficient drainage system is needed to cope with saline soils and irrigation water. New methods of irrigation, such as under-tree sprinklers and trickle irrigation, and new methods of using existing systems, such as using overhead sprinklers at night to reduce evaporation, are all being used (Crabb 1970), but they do nothing about the real difficulties (Maunsell & Partners 1979).

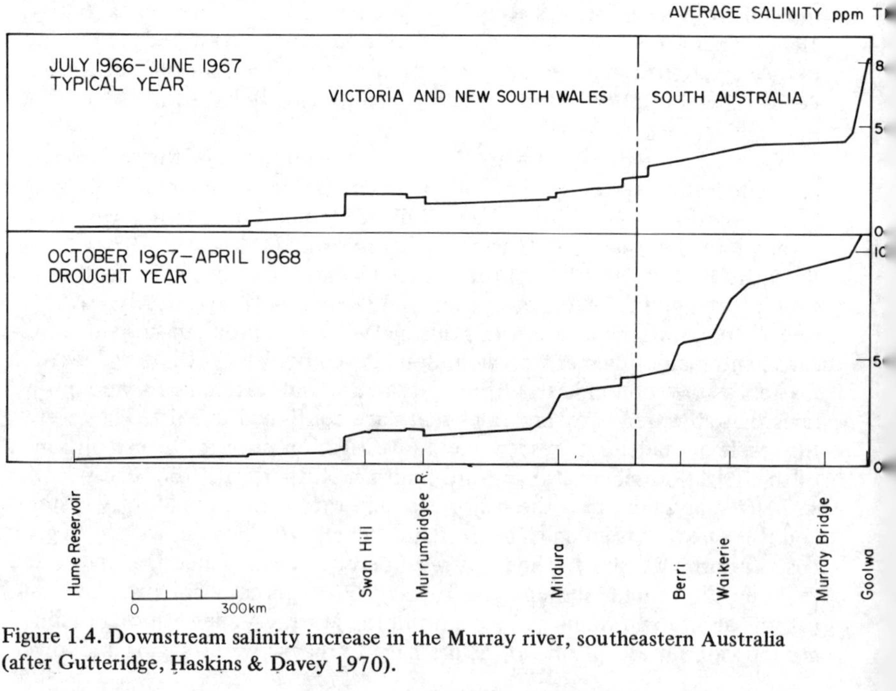

Figure 1.4. Downstream salinity increase in the Murray river, southeastern Australia (after Gutteridge, Haskins & Davey 1970).

1.5 UNDERSTANDING THE PROBLEMS

It will be clear that many of the problems associated with irrigation agriculture in the semi-arid zone are the result of a lack of knowledge and understanding of the areas being irrigated. In socio-economic terms, the costs have been and continue to be high, for individual farmers, irrigation areas, and nations. Further, if such continuing problems are to be remedied or at least alleviated, then the lack of knowledge will have to be remedied. This will involve much more than simply mapping geology, soils, and geomorphology: it will require an understanding of the depositional history of each area (Kovda 1973). For example, a study of surface and subsurface soil salinity in the West Central Gezira of the Sudan suggested that the 'depositional history is the dominant control, acting through soil texture, permeability, and topography' (Williams 1968). The conditions under which the soils and groundwater were deposited fluctuated widely, and in most instances were different from those prevailing today. An understanding of the fact that soils and groundwater deposits are out of phase with current conditions is essential to efficient irrigated agriculture in many areas. This is perhaps especially true where groundwater is being used, from both rechargeable and non-rechargeable aquifers (Williamson 1964). In the former case, the depositional history aids the understanding of the movement of water in the alluvial plains and the recharge of the aquifers. In the latter case, use is being made of a non-renewable resource – it is the 'mining' of 'fossil' waters – which cannot be recharged (Faure & Williams 1977). Such a water source is hardly an adequate basis for an irrigation system.

Detailed studies are leading to a greater understanding of salinity and water supply problems in many irrigated areas, such as the Riverine Plan (Butler 1974) and the Namoi Valley in Australia. This is also true of salinity problems in non-irrigated areas (Conacher 1979; Northcote & Skene 1972). In time, there is no doubt that the increased knowledge will lead to more efficient water use and higher agricultural productivity. In the semi-arid zone, with its limited water and soil resources, there is clearly much to be gained from increased knowledge of the areas being irrigated.

1.6 REFERENCES

Amiran, D.H.K. 1966, Man in arid lands. In E.S. Hills:219-254.

Anon 1976, Salinity control and drainage – a strategy for the Northern Plains. *Aqua: official journal, State Rivers & Water Supply Commission of Victoria,* Summer-autumn: 6-9.

Batisse, M. 1969, Problems facing arid-land nations. In W.G. McGinnies & B.J. Goldman (eds), *Arid lands in perspective.* Tucson, American Ass. for the Advancement of Science, Washington/University of Arizona Press:3-12.

Batisse, M. 1976, Water needs and resources. Rational management: the key to the supply-demand equation. *Nature and resources* 12(4):2-9.

Butler, B.E. 1974, Soils. In Frith & Sawer:13-28.

Butler, B.E. et al. 1973, *A geomorphic map of the riverine plain of south-eastern Australia.* Camberra, ANU Press.

Conacher, A. 1979, Salinity problems in Southwestern Australia. *Search* 19(5):162-164.

Cooke, R.U. & A. Warren 1973, *Geomorphology in deserts.* London, B.T. Batsford Ltd.

Cornejo T. A. 1974, Resources of arid South America. In H.E. Dregne:345-380.

Crabb, P. 1969, Some aspects of salinity in the South Australian Upper Murray: some observations. *Austral. Geogr.* 11(2):170-172.

Crabb, P. 1970, Aspects of the South Australian citrus industry. *Scott. Geogr. Mag.* 86(1): 41-51.

Davidson, B.R. 1974, Irrigation economics. In Frith & Sawer:193-212.

Downing, T.E. & M. Gibson (eds) 1974, *Irrigation's impact on society.* Anthropol. Papers of the Univ. of Arizona 25. Tucson, University of Arizona Press.

Dregne, H.E. (ed.) 1974, *Arid lands in transition.* Washington, Amer. Ass. for the Advanc. of Sci.

Evenari, M. et al. 1971, *The Negev: the challenge of a desert.* Cambridge, Harvard University Press.

Faure, H. & M.A.J. Williams 1977, Quaternary palaeoclimates and water resources. *Bull. de l'ASEQUA* 51:75-81.

Frith, H.J. & G. Sawer 1974, *The Murray waters: man, nature and a river system.* Sydney, Angus & Robertson Ltd.

Goodall, D.W. & R.A. Perry (eds) 1979, *Arid-land ecosystems: structure, functioning and management.* 1. Intern. Biol. Progr. 16. Cambridge University Press.

Gutteridge, Haskins & Davey 1970, *Rivery Murray salinity investigations,* 1 and 2. Canberra, River Murray Commission.

Hernandez X.E. 1974, Mexican experience. In H.E. Dregne: 317-343.

Hills, E.S. (ed.) 1966, *Arid lands: a geographical appraisal.* London, Methuen & Co. Ltd, UNESCO, Paris.

Jennings, J.N. 1955, Cyclic salt in Australia. *Geogr. Rev.* 45(2):277-279.

Jennings, J.N. 1956, The salinity problem in arid lands. *Geogr. Rev.* 46(3):423-425.

Jewitt, T.N. 1966, Soils of arid lands. In E.S. Hills:103-125.

Kovda, V.A. 1973, Landscapes in relation to irrigation, drainage and salinity. In V.A. Kovda et al. (ed.), *Irrigation, drainage and salinity: an international source book.* London, UNESCO/FAO/Hutchinson & Co. Ltd.:155-176.

Langford-Smith, T. & J. Rutherford 1966, *Water and land: two case studies in irrigation.* Canberra, ANU Press.

McMichael, D.F. et al. 1976, *River Murray working party: report to Steering Committee of Ministers, October 1975.* Canberra, Austral. Govt. Publ. Serv.

Maunsell & Partners 1979, *Murray valley salinity and drainage: development of a coordinated plan of action.* Canberra, Maunsell & Partners Ltd.

Meigs, P. 1953, World distribution of arid and semi-arid homoclimates. *Arid Zone Res.* 1:203-210.

Morris, A.S. 1969, The development of the irrigation economy of Mendoza, Argentina. *Annals of the Ass. of Amer. Geogr.* 59(1):97-115.

National Academy of Sciences 1968, *Water and choice in the Colorado Basin: an example of alternatives in water management.* Washington, Nat. Acad. of Sci.

National Council for Research 1975, *Agricultural research capabilities and scope in the Sudan.* Khartoum, Nat. Council for Res., Agric. Res. Council.

Newhouse, F. et al. 1950, *Irrigation in Egypt and the Sudan, the Tigris and Euphrates Basin, India and Pakistan.* London, Longmans, Green & Co. Ltd.

Nir, D. 1974, *The semi-arid world: man on the fringe of the desert.* London, Longman Group Ltd.

Northcote, K.H. & J.K.M. Skene 1972, *Australian soils with saline and sodic properties.* Soil Publ. 27. Melbourne, CSIRO.

Petrov, M.P. 1974, Deserts of Asia. In H.E. Dregne:279-302.
Senate Standing Committee on Science and the Environment 1979, *Continuing scrutiny of pollution – the River Murray.* Progress Report June 1979. Parliamentary Paper, 117. Canberra, Commonwealth Govt. Printer.
Vermeer, E.B. 1977, *Water conservancy and irrigation in China: social, economic and agro-technical aspects.* The Hague, Leiden University Press.
Williams, M.A.J. 1968, Soil salinity in the West Central Gezira, Republic of the Sudan. *Soil Sci.* 105(6):451-464.
Williamson, W.H. 1964, The development of groundwater resources of alluvial formations. In *Water resources use and management.* Melbourne University Press:195-211.
Worthington, E.B. (ed.) 1977, *Arid land irrigation in developing countries: environmental problems and effects.* Oxford, Pergamon Press Ltd.

MARTIN A.J.WILLIAMS
Macquarie University, North Ryde, Australia

2

Quaternary environments in Northern Africa

2.1 INTRODUCTION

The key problem confronting the tenacious inhabitants of the vast area of desert and semi-desert which extends across 10,000 km from the Atlantic littoral of the western Sahara to the salt lakes of Rajasthan in northern India is lack of water. The outcome of capricious rainfall and high evaporation is a sparse plant cover and recurrent drought. In this context rain-fed cultivation is hazardous, and opportunistic use of ephemeral grazing by nomadic pastoralists is often the only means to wrest a living from the harsh environment.

The response of plants to the low, erratic rainfall may vary considerably from one soil to another. In his pioneering study of the trees of the Sudan, Smith (1949) demonstrated that trees requiring a given rainfall when growing on clay soils required only two-thirds of that amount when growing on sands. An opposite response is shown by Gezira cotton, the yield of which increases as the clay content of the soil increases, which in part accounts for the importance of irrigated cotton on the clay plains of the central Sudan (Finck 1960, 1963). Among the reasons for this differential response to clay content is the ability of *Acacia* trees to tolerate the lower nutrient status of sandy soils, coupled with low rates of transpiration from their leaves.

High evaporation and minimal leaching of the soil result in progressive accumulation of carbonate, sulphate and chloride salts within the profile. The least soluble salts are precipitated first, so that with increasing concentration of the soil- and ground-water ions, there is a sequential precipitation of calcium carbonate, gypsum (calcium sulphate) and halite (sodium chloride). The greater the content of soluble and exchangeable sodium within the soil, the more difficult is seed germination and plant growth. And so a further problem afflicts the cultivator of semi-arid lands: the adverse influence of saline and alkaline soils.

Removal of the salts by leaching with irrigation water may alleviate the problem, but if the water is itself slightly saline and if the infiltration capacity of the soil is low, the remedy may prove worse than the ill, as illustrated by Crabb in Chapter 1. The twin issues are therefore soil quality and water quality, both of which are bound up with regional geological history, and in particular with the depositional and climatic oscillations of the last few million years.

Before discussing the regional geology and stratigraphy of the central Sudan in Chapters 4 and 5, and the relationship between Nile history and evolution of the central Sudanese landscape in Chapters 6 to 10, it seems useful to consider the Sudan within the broader context of Africa north of the equator. The Quaternary history of northern Africa is the subject of this chapter, that of the Sudan itself follows in Chapter 3.

2.2 DROUGHT PREDICTION AND CLIMATIC CHANGE

The Sahara desert is the dominant geographical feature in northern Africa. Certainly the largest, it is also one of the hottest and driest deserts in the world. During times of drought it is oft asserted that the desert is advancing – an advance variously attributed to overgrazing, cultivation, and drought-induced death of the plants responsible for stabilising the otherwise shifting sands of the desert margin. Future survival of those who dwell in the broad transitional zone between desert and southern savanna or northern steppe depends upon the ability of the plant cover to re-establish itself after drought. The causes of ecological deterioration and of soil erosion urgently need to be understood and checked, and considerable thought is now being given to methods of long-term management of arid lands (Adams 1975; Mabbutt 1976, 1978).

As in other long-term predictions of climatic change, there are two main approaches to forecasting future droughts. One, theoretical, tries to predict future change by modelling the effect of slight changes in boundary conditions (especially surface reflectivity, or albedo) upon global atmospheric circulation. The other, historical, seeks to document the pattern of past climatic changes at different scales in space and time (Williams 1979), and to extrapolate these into the future. Climate is of course a mental construct, and former climates can only be deduced at second-hand from the geological and biological evidence thought to be diagnostic of such climates – a procedure that often seems unduly intuitive. Livingstone (1975) and Flint (1976) have reviewed this problem with lucid good sense.

As a first step in understanding the evolution of the present North African landscape, we need to know when and why the Sahara first became a desert (Williams & Faure 1980).

2.3 ORIGIN OF THE SAHARA

During the Triassic Africa was part of the great continent of Gondwana, which also included Australia, Antarctica, India and South America. By Jurassic times West Gondwana, comprising Africa and South America, had separated from East Gondwana, comprising Australia, Antarctica and India. Separation of each of the constituent continental segments of former Gondwana began during the Cretaceous. A slow clockwise rotation and northward movement of the African plate took place during the ensuing 75 million years of Cenozoic time (Habicht 1979).

One result of these movements was a change in the position of Africa relative to the equator. During the Cretaceous Africa was positioned such that the equator ran from S.W. Nigeria through central Chad into N.W. Sudan and Saudi Arabia (Habicht 1979). By early Tertiary time, Africa was only a few degrees south of its present position. During the Miocene and Pliocene a slight clockwise rotation brought Africa into contact with Europe. The resultant crustal deformation was accompanied by considerable uplift in the Atlas, and by volcanism and updoming in the Hoggar, Tibesti, Aïr and Jebel Marra mountains (Vincent 1960; Rognon 1967; Girod 1971; Vail 1972).

Another outcome of the slow northward drift of Africa during post-Palaeozoic time was a progressive southward shift of the equatorial forest belt which once ran diagonally across the Sahara from SW to NE. Northwestern Africa was in dry tropical latitudes well before Egypt and Sudan, a situation responsible for the abundant evaporite formations within the Mesozoic and Tertiary rocks of Tunisia and Algeria (Coque 1962; Conrad 1968).

Changes in the Tertiary flora of the Sahara (Maley 1980a) stemming from its northward drift into latitudes characterised by dry subsiding air coincided with fairly sudden post-Eocene oceanic cooling in high latitudes, and with the progressive growth of continental ice caps in both hemispheres. Glaciers were probably present on Antarctica early in the Oligocene, and a large ice cap was well established there by 10 m.y. ago (Shackleton & Kennett 1975). Continental ice sheets were forming in northern high latitudes over 3 m.y. ago, and there was a significant increase in the total volume of ice towards 2.5 m.y. ago (Shackleton & Opdyke 1977).

Associated with the late Tertiary build-up of ice in high latitudes were major changes in the vegetation of northern Africa. During the early Tertiary (Palaeocene and Eocene) conditions in the southern sector of the present Sahara were dominantly equatorial, and widespread deep weathering was characteristic of this time (Faure 1966; Greigert & Pougnet 1967). At the onset of the Oligocene sea surface temperatures fell sharply. From Oligocene to Miocene time Sudano-Guinean woodland and Sudanian savanna woodland covered much of what is now the Sahara (Maley 1980a). Pollen spectra from scattered localities in northern Africa show that many elements of the present Saharan flora were already present during the Pliocene (Maley 1980a) suggesting that replacement of tropical woodland by plants adapted to aridity was in progress during the late Miocene and early Pliocene.

Another indication of increasing aridity towards the end of the Tertiary is the first appearance of wind-blown quartz sand interstratified among Plio-Pleistocene sediments in the Chad basin (Servant 1973; Servant & Servant-Vildary 1980). It is perhaps significant that the lacustrine diatom flora points to temperatures cooler than those prevalent in this region today at about this time. The equation of drier conditions in the Chad basin with lower temperatures is of more than local palaeoclimatic interest.

Theoretical arguments suggest that as roughly 70 % of world precipitation now falls between latitudes 40° N and 40° S, a reduction in evaporation from the oceans within this zone would reduce total world precipitation. Estimates vary, but a recent reappraisal by Flohn suggests that during the last glacial maximum when sea-surface temperatures were lower (McIntyre et al. 1976) total evaporation was reduced by 18 %, and that between latitudes 10° N and 10° S by nearly 70 % (Flohn & Nicholson 1980).

Confirmation of these arguments is offered by the isotopic, mineralogical and micro-fossil composition of Pleistocene marine sediments from the Red Sea and Atlantic borders of northern Africa. The oxygen isotopic composition of planktonic foraminifera collected from the Gulf of Aden and Red Sea shows that during the past 0.25 m.y., at least, glacial maxima were times of extreme aridity and very high evaporation from the Red Sea (Deuser et al. 1976). In the equatorial Atlantic those times during the last 0.6 m.y. when there was a major input of desert dust into the ocean were also times of low sea-surface temperature (Parmenter & Folger 1974; Bowles 1975). The dust included fine quartz, opal phytoliths, and fossil freshwater diatoms from the floors of former lakes. The *harmattan* dust which is blown from the Chad basin across Nigeria and out into the Atlantic today has a very similar composition to that found in the Pleistocene oceanic cores (G.McTainsh, personal communication).

The late Tertiary desiccation which brought into being the world's greatest desert was the outcome of several factors. Northward movement of the African plate during the past hundred million years transported northern Africa from moist equatorial latitudes into the dry sub-tropical zone. Late Tertiary build-up of continental ice sheets and related cooling of the oceans caused a global reduction in evaporation from the surface of the oceans, and a corresponding decrease in world precipitation. This effect was especially marked within the intertropical zone.

The consequences of these changes were predictable and dramatic. Major rivers which had previously eroded large valleys dwindled and dried up. Valleys became choked with alluvium and with colluvium washed off the previously vegetated and stable slopes. Wind, no longer impeded by vegetation, mobilised the alluvium and fashioned some of it into dunes. Silt- and clay-sized particles were blown from the land out to sea. Some of the desert dust was trapped by vegetation along the semi-arid fringes of the desert to form extensive mantles of fine-textured soil material. The timing of these events no doubt varied from one region to another, but few of them are accurately dated.

In the now arid Hoggar mountains of southern Algeria there seem to have been several phases during the late Pliocene or early Pleistocene when the mountain valleys contained tropical freshwater lakes, which subsequently dried out (Rognon 1967). The same seems true of the Tibesti further south (Maley et al. 1970). In western Sudan lake diatomites of probable early Pleistocene age occupy a lava-dammed valley northwest of Jebel Marra (Williams et al. 1980), and in the Afar desert of Ethiopia and Djibouti lakes occupied now arid fault-troughs on a number of occasions during the late Pliocene and early Pleistocene (Gasse 1975; Gasse et al., in press).

2.4 QUATERNARY CLIMATIC FLUCTUATIONS

Two main problems impede interpretation of changes in the Quaternary landforms and biota in northern Africa and elsewhere. One is the fragmentary nature of the terrestrial record; the other is the inherently ambiguous character of that record.

It is rare for countries in the recently deglaciated middle latitudes of Eurasia and the Americas to retain evidence of more than four or five glacial depositional episodes. In contrast to this is the evidence of at least 17 glacial-interglacial cycles in the past 1.75 m. years preserved in deep-sea core deposits and in loess-mantles in unglaciated areas (Kukla 1977; Fink & Kukla 1977). As noted earlier, the onset of significant mid-latitude glaciation was about 2.5 m.y. ago in both hemispheres (Shackleton & Opdyke 1977), so that the total number of Plio-Pleistocene glacial cycles may have been closer to 25 than 17. The significance for northern Africa is clear. As we have seen, peak aridity in the dry sub-tropics of Africa coincided with times of global temperature minima, that is to say, with times of maximum glaciation (Fairbridge 1970; Parmenter & Folger 1974).

The pattern of intertropical glacial aridity outlined above is not only true for Africa (Street & Grove 1976; Livingstone 1975, 1980), but also affected India-Pakistan, the Middle East, Australia and South America (Fairbridge 1970; Williams 1975), with corresponding impacts upon plant cover, runoff and sediment yield in major river basins like the Nile (Adamson et al. 1980; see also Chapter 10).

The second problem is the equivocal nature of the Quaternary stratigraphic record in northern Africa. Particularly in the East African rifts, but equally in and near the Atlas Mountains, depositional changes may reflect local volcano-tectonic activity rather than, or in addition to, late Cenozoic climatic oscillations of global extent. It is clearly important to be able to distinguish depositional and biological changes arising from continental fluctuations in climate from those resulting from volcanism, rifting or uplift, for the latter can accentuate or attenuate the impact of climatic change. For instance, it is often difficult to decide whether a lake owes its origin and eventual demise to crustal movements, to lava flows, or to climatic change. A solution is to compare lake level fluctuations with the fossil pollen and diatom record, sadly all too often incomplete.

Pollen spectra from fluvio-deltaic sediments of the Omo River in southern Ethiopia reveal a decrease in rainfall between 3.0 and 2.0 m.y. ago, leading to extinction of the *Antrocaryan* flora and culminating in a drier-than-present climatic phase between 2.35 and 2.10 m.y. ago (Bonnefille 1979). Further afield, in northern Kenya, pollen spectra point to a dry episode towards 1.4-1.5 m.y. ago (Vincens 1979) and somewhat earlier at Olduvai in Tanzania there was an arid interval towards 1.7 m.y. ago (Bonnefille 1979).

Pollen spectra from sporadic samples such as these offer no real insight into the long-term climatic oscillations responsible for these biological changes. What is needed is a long, continuous fossiliferous sequence, of which northern Africa has all too few.

A partial exception is the thick sequence of lake deposits at 2,300 m elevation in the plain of Gadeb, south-central Ethiopia. Detailed analyses by Gasse of the relative proportions of planktonic and littoral diatoms revealed four climatically-controlled transgressive/regressive cycles between 2.71 and 2.35 m.y. ago (Williams et al. 1979; Gasse 1980). Once the abundant pollen grains within the Gadeb diatomites have been analysed, there should be a very complete record of regional changes in the late Pliocene flora in this part of northern Africa.

Three important conclusions may be drawn from the sparse and still inadequately studied Quaternary record of northern Africa.

1. The late Pliocene trend towards drier and/or colder conditions evident in Tunisia (Coque 1962), Algeria (Conrad 1968), Chad (Servant & Servant-Vildary 1980), Ethiopia (Bonnefille 1979) and Kenya (Bonnefille 1976; Cerling et al. 1977) was related to world-wide changes in ice volume, ocean currents and sea-surface temperatures (Faure 1980; Schnitker 1980). The same trend is also evident in Australia, where the history of late Cenozoic desiccation roughly parallels that of the Sahara.

2. Each glacial-interglacial cycle appears to have lasted roughly 90,000-100,000 years (Fink & Kukla 1977), with a slow build-up towards full-glacial conditions, followed by a short, sharp deglaciation of about 10,000 years' duration (Broecker & Van Donk 1970). The four transgressive cycles evident in the late Pliocene Ethiopian Lake Gadeb are consistent with this pattern, but the precise relationship between lake level, temperature, precipitation and runoff is not yet known.

3. The climax to a middle latitude glacial phase seems to be associated with peak aridity in low latitudes, as shown by the drying out of Lake Chad (Servant 1973) and of Lake Abhe in Ethiopia (Gasse 1977) between about 18,000 and 12,000 years ago.

A corollary to these three conclusions is that the alternation from wet to dry climates, or from fluviatile deposition to aeolian activity, will lead to the progressive spread of sand dunes and to the progressive disintegration of drainage networks in northern Africa during the Quaternary, as documented by Coque (1962) and Conrad (1968) in Tunisia and Algeria, as well as by Al Sayari and Zötl (1978) in Saudi Arabia. The terminal Pleistocene phase of dune reactivation along the southern margin of the Sahara (Talbot 1980; see also Figure 8 in Vail 1978) was but one of a score or so of Quaternary arid periods, each of which, by killing off certain elements of the plant cover, accentuated the trend towards desert expansion. The surprisingly youthful ages of the Erg Chech and the Great Western Erg (Conrad 1968) are easier to understand, given a context of intermittent but irreversible drainage disintegration and devegetation.

It is currently possible to reconstruct the late Quaternary fluctuations in lake level or river regime or plant cover for about a dozen well-dated localities in northern Africa (Williams in press). Climatic interpretation of these changes at other than a general level (Rognon & Williams 1977; Nicholson & Flohn

1980) is hazardous, although such attempts do help focus attention on potential research problems. For northern Africa as a whole, the interval from ⩾ 40,000-20,000 B.P. was one of groundwater recharge, active runoff and high lake levels, with a renewal of such conditions, on a more moderate scale, in the southern Sahara during the first half of the Holocene; from 11,000-12,000 B.P. to about 7,500 B.P.

The moist interval towards 6,500-4,500 B.P., which is evident on both margins of the Sahara as well as in now hyper-arid parts of Libya, Egypt and the Sudan, saw a vast expansion of Neolithic pastoralists throughout the deserts of northern Africa, followed by a great exodus to the less arid margins and to the Nile valley after about 4,500 B.P. (Clark 1980). Two little is known of the migrations of the proto-historic Libyco-Berber peoples and of other possible builders of the enigmatic stone monuments that are a feature of the late Holocene Sahara (Milburn & Nowak 1978-79) to warrant comment here. Both Maley (1980b) and Nicholson (1980) have put a convincing case for less arid conditions in the Sahel during at least part of the Little Ice Age of c. 1450 A.D. to 1850 A.D.

The natural tendency of nomadic Sahel dwellers is to move north into the desert when good rains and available pastures allow it, and to retreat south when drought strikes, as in 1913, the early 1940's and again in 1968-73. If intervals between droughts are too short to allow the plant cover to regenerate, and if pressure of numbers makes free movement more difficult, these age old coping strategies may fail, with known and tragic results.

2.5 REFERENCES

Adams, M.E. 1975, A development plan for semi-arid areas in western Sudan. *Expl. Agric.* 11:277-297.

Adamson, D.A., F.Gasse, F.A.Street & M.A.J.Williams 1980, Late Quaternary history of the Nile. *Nature* 288:50-55.

Al-Sayari, S.A. & J.G.Jötl 1978, *Quaternary Period in Saudi Arabia,* vol. 1. Wien, Springer-Verlag: 334.

Bonnefille, R. 1976, Implications of pollen assemblage from the Koobi Fora Formation, East Rudolf, Kenya, *Nature* 264:403-407.

Bonnefille, R. 1979, Méthode palynologique et reconstitution paléoclimatiques au Cénozoique dans le Rift Est Africain. *Bull. Soc. géol. Fr.* (7) XXI(3):331-342.

Bowles, F.A. 1975, Palaeoclimatic significance of quartz/illite variations in cores from the eastern equatorial North Atlantic. *Quat. Res.* 5:225-35.

Broecker, W.S. & J.Van Donk 1970, Insolation changes, ice volumes, and the O^{18} record in deep-sea cores. *Reviews of Geophysics and Space Physics* 8:169-198.

Cerling, T.E., R.L.Hay & J.R.O'Neil 1977, Isotopic evidence for dramatic climatic changes in East Africa during the Pleistocene. *Nature* 267:137-138.

Clark, J.D. 1980, Human populations and cultural adaptations in the Sahara and Nile during prehistoric times. In: M.A.J.Williams & H.Faure (eds.), *The Sahara and the Nile:*527-582. Rotterdam, Balkema.

Conrad, G. 1968, *Evolution continental post-hercynienne du Sahara algérien.* Paris: CNRS: 527.

Coque, R. 1962, La Tunisie présaharienne. Etude géomorphologique. Paris, Armand Colin:476.

Deuser, W.G., E.H.Ross & L.S.Waterman 1976, Glacial and pluvial periods: their relationship revealed by Pleistocene sediments of the Red Sea and Gulf of Aden. *Science* 191:1168-70.

Fairbridge, R.W. 1970, World paleoclimatology of the Quaternary. *Rev. Géogr. Phys. Géol. Dyn.* (2): XII(2):97-104.

Faure, H. 1966, Reconnaissance géologique des formations sédimentaires post-paléozoiques du Niger oriental. *Mém. Bur. Rech. Géol. Min.* 47:630.

Faure, H. 1980, Le cadre chronologique des phases pluviales et glaciaires de l'Afrique. II. In: J.Ki-Zerbo (ed.), *Histoire Générale de l'Afrique* vol. I:409-434. Paris, Unesco.

Finck, A. 1960, Re-examination of the sodium value concept for soil evaluation. Gezira Res. Station, Wad Medani, Annual Report 1959/60.

Finck, A. 1963, *Tropische Böden.* Parey, Hamburg.

Fink, J. & G.Kukla 1977, Pleistocene climates in Central Europe: At least 17 Interglacials after the Olduvai Event. *Quat. Res.* 7:363-371.

Flint, R.F. 1976, Physical evidence of Quaternary climatic change. *Quat. Res.* 6:519-528.

Flohn, H. & S.Nicholson 1980, Climatic fluctuations in the arid belt of the 'Old World' since the last Glacial maximum; possible causes and future implications. *Palaeoecology of Africa* 12:3-22.

Gasse, F. 1975, *L'évolution des lacs de l'Afar Central (Ethiopie et TFAI) du Plio-Pléistocène à l'Actuel: Reconstitution des palaéomilieux lacustres à partir de l'étude des diatomées.* D.Sc. thesis, 3 vols:406, Univ. of Paris.

Gasse, F. 1977, Evolution of Lake Abhé (Ethiopia and TFAI), from 70 000 B.P. *Nature* 265:42-45.

Gasse, F. 1980, Les diatomées lacustres plio-pléistocenes du Gadeb, Ethiopie. *Rev. Algologique,* spec. issue.

Gasse, F., O.Richard, D.Robbe, P.Rognon & M.A.J.Williams (in press), Evolution tectonique et climatique de l'Afar Central d'après l'étude des sédiments plio-pléistocenes. *Bull. Soc. Géol. Fr.*

Girod, M. 1971, *Le massif volcanique de l'Atakor (Hoggar, Sahara algérien). Etude pétrographique, structurale et volcanologique:* 158. Paris, CNRS.

Greigert, J. & R.Pougnet 1967, Essai de description des formations géologiques de la République du Niger. *Mém. Bur. Rech. Géol. Min.* 48:236

Habicht, J.K.A. 1979, *Paleoclimate, paleomagnetism, and continental drift.* Am. Assoc. Petrol. Geol. Studies in Geology No. 9, Tulsa, Oklahoma.

Kukla, G. 1977, Pleistocene land-sea correlations. I. Europe. *Earth-Science Rev.* 13:307-374.

Livingstone, D.A. 1975, Late Quaternary climatic change in Africa. *Ann. Rev. Ecol. Systematics* 6:249-280.

Livingstone, D.A. 1980, Environmental changes in the Nile headwaters. In: M.A.J. Williams & H.Faure (eds.), *The Sahara and the Nile:*339-359. Rotterdam, Balkema.

Mabbutt, J.A. (ed.) 1976, *Problems in the development and conservation of desert and semidesert lands.* International Geographical Union, Working Group on Desertification, 23rd I.G.C., Ashkhabad, USSR. Australia, Univ. of New South Wales.

Mabbutt, J.A. (ed.) 1979, *Proceedings of the Khartoum Workshop on Arid Lands Management, October 1978.* Tokyo, United Nations University.

Maley, J. 1980a, Les changements climatiques de la fin du Tertiaire en Afrique: leur conséquence sur l'apparition du Sahara et de sa végétation. In: M.A.J.Williams & H.Faure (eds.), *The Sahara and the Nile:*63-86. Rotterdam, Balkema.

Maley, J. 1980b, *Etudes palynologiques dans le bassin du Tchad et paléoclimatologie de*

l'Afrique Nord-tropicale de 30.000 ans à l'époque actuelle. Thesis:586, Montpellier.
Maley, J., J.Cohen, H.Faure, P.Rognon & P.M.Vincent 1970, Quelques formations lacustres et fluviatiles associées à différentes phases du volcanisme au Tibesti (Nord du Tchad). *Cahier ORSTOM sér. Géol.* 2:127-152.
McIntyre, A. & CLIMAP project members 1976, The surface of the Ice-age earth. *Science* 191:1131-44.
Milburn, M. & H.Nowak 1978-79, On the study of Libyan and Saharan stone structures as a possible means of researching early desert trade and contacts. *Almogaren,* (Graz) 9-10:107-134.
Nicholson, S.E. 1980, Saharan climates in historic times. In: M.A.J.Williams & H.Faure (eds.), *The Sahara and the Nile:*173-200. Rotterdam, Balkema.
Nicholson, S.E. & H.Flohn 1980, African environmental and climatic changes and the general atmospheric circulation in Late Pleistocene and Holocene. *Climatic Change* 2:313-348.
Parmenter, C. & D.W.Folger 1974, Eolian biogenic detritus in deep sea sediments: a possible index of equatorial Ice Age aridity. *Science* 185:695-8.
Rognon, P. 1967, *Le massif de l'Atakor et ses bordures (Sahara central). Etude géomorphologique:*555. Paris, CNRS.
Rognon, P. & M.A.J.Williams 1977, Late Quaternary climatic change in Australia and North Africa: A preliminary interpretation. *Palaeogeogr., Palaeoclimat., Palaeoecol.* 21:285-327.
Schnitker, D. 1980, Global paleoceanography and its deep water linkage to the Antarctic glaciation. *Earth-Science Rev.* 16:1-20.
Servant, M. 1973, *Séquences continentales et variations climatiques: évolution du bassin du Tchad au Cénozoique supérieur.* Thesis:368, Univ. of Paris.
Servant, M. & S.Servant-Vildary 1980, L'environnement quaternaire du bassin du Tchad. In: M.A.J.Williams & H.Faure (eds.), *The Sahara and the Nile:*133-162. Rotterdam, Balkema.
Shackleton, N.J. & N.D.Opdyke 1977, Oxygen isotope and palaeomagnetic evidence for early Northern Hemisphere glaciation. *Nature* 270:216-219.
Smith, J. 1949, Distribution of tree species in the Sudan in relation to rainfall and soil texture. *Sudan Govt. Min. Agric. Bull. No. 4.*
Street, F.A. & A.T.Grove 1976, Late Quaternary lake level fluctuations in Africa: environmental and climatic implications. *Nature* 261:385-390.
Talbot, M.R. 1980, Environmental responses to climatic change in the West African Sahel over the past 20,000 years. In: M.A.J.Williams & H.Faure (eds.), *The Sahara and the Nile:*37-62. Rotterdam, Balkema.
Vail, J.R. 1972, Jebel Marra, a dormant volcano in Darfur Province, Western Sudan. *Bull. Volc.* 36:251-65.
Vail, J.R. 1978, *Outline of the geology and mineral deposits of the Democratic Republic of the Sudan and adjacent areas. Overseas Geol. & Miner. Resourc.* No. 49:68, London, HMSO.
Vincens, A. 1979, Analyse palynologique du site archéologique FxJj50, Formation de Koobi Fora, Est Turkana (Kenya). *Bull. Soc. géol. Fr.* (7) XXI(3):343-347.
Vincent, P. 1960, *Les volcans tertiaires et quaternaires du Tibesti occidental et central.* Thesis:235, Paris.
Williams, M.A.J. 1975, Late Quaternary tropical aridity synchronous in both hemispheres? *Nature* 253:617-618.
Williams, M.A.J. 1979, Droughts and long-term climatic change: recent French research in arid North Africa. *Geogr. Bull.* 11:82-96.
Williams, M.A.J. (in press), Late Quaternary prehistoric environments in the Sahara. In:

J.D.Clark & S.A.Brandt (eds.), *Causes and Consequences of Food Domestication in Africa.* Univ. of Calif. Press.

Williams, M.A.J. & H.Faure (eds.) 1980, *The Sahara and the Nile:*607, Rotterdam, Balkema.

Williams, M.A.J., D.A.Adamson, F.M.Williams, W.H.Morton & D.E.Parry 1980, Jebel Marra volcano: a link between the Nile Valley, the Sahara and Central Africa. In: M.A.J.Williams & H.Faure (eds.), *The Sahara and the Nile:*305-337. Rotterdam, Balkema.

Williams, M.A.J., F.M.Williams, F.Gasse, G.H.Curtis & D.A.Adamson 1979, Plio-Pleistocene environments at Gadeb prehistoric site, Ethiopia. *Nature* 282:29-33.

GERALD E.WICKENS
Royal Botanic Gardens, Kew, UK

3

Palaeobotanical speculations and Quaternary environments in the Sudan

3.1 INTRODUCTION

That there have been significant changes in effective precipitation, runoff and plant cover in the Sudan during the past 40 000 years and more is no longer in doubt (Wickens 1975; Williams & Adamson 1976). That such changes must have had a significant influence upon early man is equally certain. But the magnitude and frequency of these late Quaternary environmental changes, although known with increasing accuracy for certain regions like the Afar and Chad (Chapter 2), remain more elusive when we consider the Sudan. This chapter discusses the evidence for Quaternary environmental changes in the Sudan as a whole, and presents a series of very provisional and hypothetical reconstructions of the vegetation of the Sudan at selected periods during the late Quaternary.

3.2 EVIDENCE FOR ENVIRONMENTAL CHANGE DURING THE LATE QUATERNARY

Inductive attempts to reconstruct late Quaternary environments in Africa are based upon evidence from geology, biology, and archaeology. Much of the evidence is fragmentary and sporadic in both time and space. It is no surprise that conclusions based upon such data are often uncertain and sometimes contradictory. The geological evidence includes fixed dunes in now humid areas and former lake deposits in now arid areas, and has been evaluated in Chapter 2. Our present concern is principally with the biological evidence, including fossil vertebrate, invertebrate and plant remains, and relict plant communities. The archaeological evidence is only touched upon briefly here, since it will be considered at length in another volume.

3.2.1 *Geological evidence*

Geological evidence of environmental change should consist ideally of absolutely dated stratigraphic sections comprising fossil-bearing fluviatile, lacustrine and aeolian deposits which can be traced laterally and correlated over large areas. Problems involved in interpreting even the more widespread deposits are discussed by Vail in Chapter 4. At present there is a growing body of increasingly well dated evidence from central and northern Sudan, ranging from shell-bearing lake sediments, evaporites, palaeo-channel sediments, and wind-blown sands (see Chapter 9). In western Sudan work is now in progress on the former shorelines of the crater lakes within the Deriba caldera at Jebel Marra, on the age of inter-dune swamp deposits of Kordofan, and on the polygenic nature of the Qoz Dango of southern Darfur (Williams et al. 1980). An early attempt to use the alignments of the fixed dunes of Kordofan to infer shifts in the direction of the dominant sand-moving winds is that of Warren (1970). His interpretation is presented here for three reasons: a) no other aeolian deposits in Sudan have been studied in such detail; b) other data from Kordofan are sparse; c) his conclusions need to be tested by future workers.

Warren (1970) studied the orientation of dune systems and prevailing wind directions in Kordofan and postulated a sequence of late Quaternary shifts in the wind and rainfall zones. Although his methods enable the southerly shifts in climatic belts to be calculated, the northerly shifts cannot, and so must be deduced by other means. He recognized four climatic sequences before the conditions prevailing today. Period I represents a very arid phase during which both wind and rainfall belts were some 450 km to the south of their present position. Sands were believed by Warren to have been moved as far south as 10°N in the region of the little-known Qozes Salsilgo and Dango, to the south of Jebel Marra. According to Parry, who recently carried out a soil survey in that area (Hunting Technical Services 1974), the soils of the Qoz Dango are probably developed on alluvial sands of predominantly local origin reworked by wind action during one or more Quaternary arid phases, a conclusion we endorse (Williams et al. 1980). Period I of Warren correlates with the Middle and Late Pleistocene arid period(s) of Grove & Warren (1968).

Period II was a very wet phase, during which there was an estimated shift in the climatic and vegetation belts to approximately 250 km north of their present position. This would give a rainfall at Khartoum approximating to that at Kosti today (487 mm). This shift is sufficient to explain the now isolated presence of such forest species as *Trema orientalis, Casearia barteri,* and *Polyscias fulva* in the gallery forests of Jebel Marra (Wickens 1971, 1976a). It is believed that the increased rainfall would have been sufficient for these species to have migrated northwards along the river valleys, from the Zaïre basin to the Jebel Marra massif. It is also sufficient to account for other known changes in the fauna and flora of northern and central Sudan. This period corresponds to group 'c' of Grove & Warren, for which they have provided dates of between 12 000 and 7 000 years BP.

Period III represents a brief dry phase, but not as dry as Period I. According to Warren there was a southward shift of over 200 km, in both wind and rainfall belts, followed by a northward retreat of the wind belts, during which the vegetation was slow to colonize the high dunes.

This was followed by Period IV, a wet phase, although less wet than

Period II, the shift in the climatic and vegetation belts being approximately 100 km north of their present position. This would probably have been insufficient for the forest species, already mentioned, to have reached Jebel Marra. However, it would have had a profound effect on the northward extension of the savanna into the Sahara. Warren (1970) suggested that the final wet phase may have been coeval with a wetter period in Chad, dated at 5 400 years BP, as well as with the period of high Nile during Neolithic times some 5 000 years ago. The climate would have enabled man to occupy the Wadi Howar area.

3.2.2 *Archaeological evidence*

The Soil Conservation Committee's report, summarized by Whyte (1950), estimated that during the late Quaternary there was a minimum northerly shift of the present-day 200 mm isohyet of about 550 km. This estimate was based on the distribution of Neolithic implements and subfossil shells scattered over a wide area of the Libyan Desert. Neolithic man had settled along the natural drainage lines within the desert, where conditions would have been most favourable.

Even today the Wadi Howar (Figure 1) has sufficient water and vegetation to support, at least for some months of the year, such animals as oryx, addax, ril, gazelle, giraffe, ostrich, hyaena, jackal, fox, red hussar monkey, antbear, porcupine, lion, and hunting dogs (Shaw 1936). The 1:250 000 Sudan Survey Map, sheet 44-M, Wadi Howar, bears the rather surprising note 'Wadi Howar much used by natives as a road from North West Darfur to El Atrun and Dongola'. Although virtually devoid of surface water today, there must have been sufficient water during Neolithic times for man to settle.

Jackson (1957), in an admirable summary of the then available archaeological and historical data, quoted Bagnold (1954), who rather surprisingly suggested that an increase in the average rainfall by as little as 24 mm would have made nomadic life possible over much of the Libyan Desert. Jackson suggested that 100 mm would appear to be the minimum for habitation away from oases and rivers, and considered Bagnold's estimate as possibly rather low.

The Conservation Committee obviously underestimated the significant response that can be obtained from quite small increases in annual precipitation. A northerly shift of the 200 mm isohyet by approximately 250 km would include the Wadi Howar area and even be sufficient to support life away from the natural drainage lines. Today permanent settlements with cultivation under similar rainfall conditions exist in northern Kordofan.

3.2.3 *Biological evidence*

The present-day distribution of biota is a consequence of the changes in flora and fauna brought about by indirect effects of major global climatic fluctuations as well as by the direct effect of environmental factors upon the biota. Plants are slow to react to environmental changes because they must rely on random dispersal mechanisms for their propagation and several successful

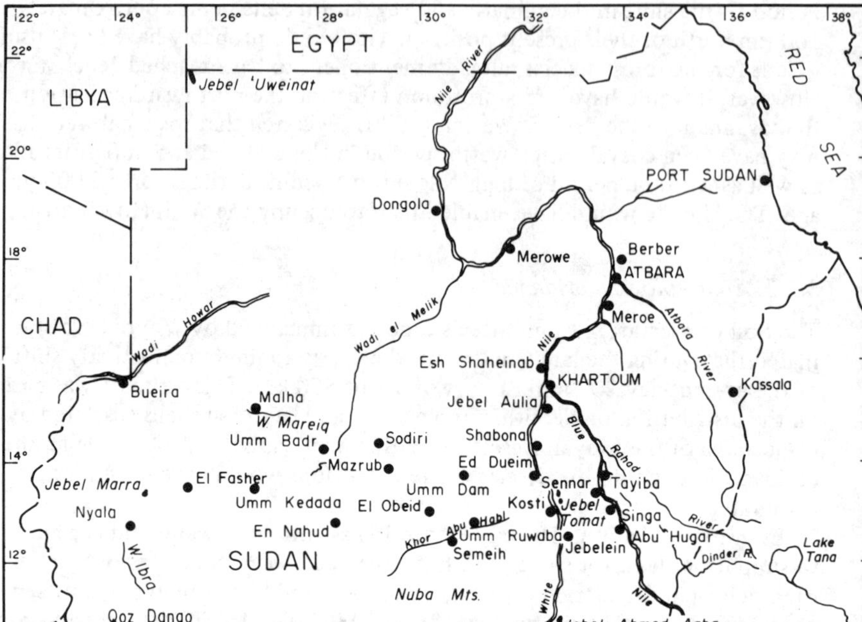

Figure 3.1. Locality map.

generations are necessary for them to establish themselves in a more favourable environment. Plant populations may be destroyed and the environment completely changed by the activities of man.

Mammals, either directly or indirectly, rely on the vegetation for their food, but because of their 'intelligence' and ability to move they are able to avoid or escape from man's sedentary activities such as cultivation, forest clearing, and drainage. They remain susceptible to elimination by hunting, the complete destruction of their habitat, or the blocking of their migration routes and the consequential adverse effect of their restricted range upon the habitat, or some other interference in the balance of nature.

These are factors which have to be considered when interpreting the fossil record. In Table I are listed the subfossil assemblies of the larger mammals found in the Nile Valley of the Sudan, together with a number of historical records from the same area.

It can be seen that the descendants of the giraffe and elephant of Early Khartoum and Esh Shaheinab were to be seen around Old Dongola in 1150 AD (Idrisi 1836). Even today they, and other game animals, are not a very great distance away but they are kept away by the density of the population and the intensity of the cultivation around the Gezira triangle. Stray elephants have been recorded from the Gezira as recently as 1920 (Grabham 1920). The hippopotamus was to be found in the Nile delta until the end of the 17th century (Cloudsley-Thompson 1967) and in the 1870's was still to be found less than 150 km to the south of Khartoum (Petherick & Petherick 1869; Schweinfurth 1873); today they have been eliminated from that stretch of the Nile by hunting.

Except for the Singa site, the subfossil mammal record does not positively indicate any major climatic change. This is partly to be expected because of the ameliorating effect of a major perennial river such as the Nile on the environment. However, there is a very strong suggestion that the fauna has either been driven from the area by the presence of man, or eliminated by hunting. Much of this change appears to have taken place during the last two centuries and is largely due, according to Cloudsley-Thompson (1967), to the population explosion and the advent of firearms.

The effect of man upon the game population is illustrated by the remarks of Baker (1867), who travelled in the upper reaches of the Rahad and reported on the wonderful fertility of the cultivation along the river banks. He also noticed that, for 200 miles from Gallabat, he had not seen a single gazelle, and he attributed the absence of game to the dense population. Grabham (1920) noted that in 1915 the same area was virtually uninhabited, with *Acacia seyal* woodland and abundant game, including elephant, lion, buffalo, and several kinds of antelope.

The fossil data help throw some light on the type of vegetation required for the various habitats of the mammals, such as swamp grasses for hippopotami, and reed swamps for the Nile lechwe. Drainage, apart from its role in agriculture, might have been used deliberately to destroy the habitats of the swamp-loving animals such as the hippopotamus and the Nile lechwe. Alternatively, the animals themselves, by sheer numbers, might have grazed out or otherwise destroyed their habitat, as is happening in a number of game reserves today where the animal populations are not controlled.

The present distribution of the Nile lechwe is restricted to the Sudd region, although it was known as far north as Khartoum some 8 000 years ago (Table 1). Its absence from Esh Shaheinab 50 km further north, by 5 500 years BP,

Table 3.1. Records of mammals in the Nile Valley and adjacent parts of Sudan (italic names come from later historical records).

Date	Location	Lat. & Long.	Rainfall Estimated mm	Present mm
c. 15 000 BP or earlier	Singa	13°11′N 33°55′E	c. 70	559
c. 8 000 BP	Khartoum	15°36′N 32°32′E	c. 490-560	163
c. 8 000 BP	Shabona	14°41′N 32°18′E	c. 790	270
5 493 ± 380 BP-5 107 ± 450 BP	Shaheinab	16°03′N 32°33′E	? 200	c. 100
c. 4 000 BP	Jebel Tomat	13°36′N 32°44′E	c. 650	c. 450
6 AD	Meroe	16°56′N 33°43′E		c. 80
1150	Old Dongola	18°05′N 30°57′E		c. 30
1770	Upper Atbara	c. 15°28′N 36°24′E		c. 325
1822	Ed Debba	18°03′N 30°57′E		c. 30
1840	Gash, nr Kassala	15°28′N 36°24′E		328
1840	Khartoum	15°36′N 32°32′E		163
1840	Aba Is.	13°20′N 32°37′E		c. 485
1852	Berber	18°01′N 33°59′E		c. 70
1860	Qoz el Merikh (Walled el Meck)	15°22′N 32°28′E		c. 160
1868	Wad Shal'ai (Wod Shellay)	14°32′N 32°13′E		c. 300
1875	Soderi	14°25′N 29°05′E		220
1920	Areidiba	15°24′N 33°41′E		560
19–	Abu Haraz	14°29′N 33°31′E		c. 400
1965	Merkhiyat Jebels	15°04′N 32°25′E		c. 165
1973	Dinder National Park	c. 13°00′N 35°00′E		c. 800

na	References	Notes
nct porcupine, extinct buffalo	Arkell, Bate, Wells & Lacaille 1951	Dry land species
Grays Waterbuck, *Buffalo, elephant,* mongoose, extinct rat, spiny field rat, Nile lechwe, *hippopotamus, rhinoce- hyaena, leopard, warthog, oribi*	Arkell 1949	Mixture of swamp and dry land species
hant, hippopotamus, giraffe, buffalo, ?Nile lechwe, *wart-* reed rat or dassie, *mongoose,* turtle, *tortoise, python*	Clark 1973	Mixture of swamp and dry land species
ard, otter, civet cat, honeybadger, *warthog, hare, gerbil, nd squirrel, oryx, gazelle, greater kudu, buffalo, giraffe, hant, rhinoceros, hippopotamus, lion, jackal, hyaena, et, roan antelope,* genet, *duika*	Arkell 1953	Dry land conditions; presence of *Limicolaria flammata* shells suggests rainfall of 400 mm
orcas) gazelle, Klipspringer, *bush pig,* cane rat, *mongoose,* upine, *hare,* hedgehog, *gerbil,* monitor lizard, genet	Clark 1973	Dry land conditions prevailing
oceros, elephant, monkeys	Pliny ex Kirwan 1957	Dry land
fe, elephant	Idrisi 1836	Dry land
hant	Bruce 1790	Dry land
	Bellefonds ex Cloudsley-Thompson 1967	
hant, rhinoceros, giraffe, hyaena, lion	Werne 1849	
lle	Werne 1849	
alo	Werne 1849	
opotamus	Churi ex Cloudsley-Thompson 1967	
opotamus	Petherick 1869	
opotamus	Schweinfurth 1893	
, hyaena	Friederichsen 1878; Wickens 1970	Savanna
hant	Grabham 1920	Savanna
hant	Jackson 1957	Savanna
mongoose, ground squirrel, jerboa, hare, ?jackal, ?hyaena, can python, Egyptian cobra, giant tortoise	Cloudsley-Thompson 1966	Savanna
hant, giraffe, buffalo, roan antelope, water buck, tiang, ter kudu, red-fronted gazelle, soemmering, gazelle, reed , bush buck, oribi, duika, Salt's dik-dik, warthog, bush lion, leopard, cheetah, hyaena, wild dog, grivet, red hussar key, baboon	Cloudsley-Thompson 1973	Savanna

however, does not denote a drier climate, but simply the absence of large areas of swamps, a factor perhaps suggestive of the damming of the White Nile lake at Khartoum during the early to middle Holocene.

Arkell (1948) reported the discovery of the fruits of *Celtis integrifolia* at Early Khartoum. The fruits were apparently gathered and used as food, as indeed they are today. According to Walter & Lieth (1967), the present-day rainfall at Khartoum is 163 mm. A northerly shift of 100 km would then give a rainfall approximating to that of Tayiba, 406 mm. Monod (1964) indicates the northern limit of present-day distribution of *Celtis integrifolia* to approximate with the 400 mm isohyet. Although apparently corroborating the deductions given above by Warren, too much significance should not be placed on this isolated conclusion because *Celtis integrifolia* is a riverine tree and consequently less likely to be affected by climatic changes. Its introduction could also have occurred earlier.

The calcareous swamp and lake deposits found in the inter-dune hollows associated with the 'High Qoz' dune formation in Kordofan (Hunting Technical Services 1964a) may also belong to this wet phase. The fossil fragments of *Phragmites* sp. collected by Hughes, Collier, and Wickens from Mazrub are from such a deposit. Associated pollen were identified as belonging to the Chenopodiaceae. Similar calcareous deposits to that at Mazrub, at En Nahud, and from near Umm Dam also require investigation.

3.2.4 *Conclusions*

Three main palaeoclimatic conclusions may be drawn from the scattered evidence reviewed here and from the more complete evidence available from the Gezira (Chapters 7 and 9).

1. During the terminal Pleistocene the climate was cool, dry and windy at least as far south as latitude 10-12°N, and was probably semi-arid throughout much of southern Sudan.

2. The early to middle Holocene was wet and warm in central Sudan and in the present savanna zone of western Sudan. Late Pleistocene dunes became submerged beneath White Nile alluvium in central Sudan, and further west small lakes and swamps occupied the depressions between the now vegetated and stabilized dunes.

3. From about middle Holocene times onwards the climate became drier, and there was a progressive shift southwards of the rainfall, vegetation and faunal zones. Man's influence upon his environment became increasingly pronounced.

3.3 MODELLING PAST VEGETATION CHANGES: A DEDUCTIVE APPROACH

3.3.1 *Assumptions involved in deductive models of vegetation change*

The approach adopted here has been succinctly described by Birks in Birks &

West (1972) as 'Quaternary palaeoecology must, by necessity, use the present to model the past'. The present-day 'climax' vegetation communities are projected into differing climatic regimes while taking into account major differences in soil texture and available soil moisture. For example, an acacia tree growing on sandy soils requires $2x$ mm annual rainfall, while on clay soils the same species requires $3x$ mm (Smith 1949).

It is assumed that these 'climax' communities have remained virtually unchanged in time. Although there is at present no botanical evidence to show otherwise, it is perhaps important to remember that a number of mammals are known to have become extinct during the past 20 000 years; should plants be the exception? It should also be remembered that progression in succession can always be expected to be slower than regression, i.e. recovery of the flora and associated fauna will be slower than the destruction of a community at the beginning of a dry period. Thus it is possible that the earlier, original climax state has been eliminated during the course of several major climatic changes. Climatic change may perhaps be regarded as occurring in waves of varying amplitudes and wave-lengths. The various maps and descriptions of former vegetation given below show what are believed to be conditions at the time of maximum aridity or humidity within a climatic period, and thus do not necessarily represent conditions over a long time within a period.

A more intuitive approach is needed regarding the effectiveness of the increased rainfall. Even a slight increase (or decrease) is likely to produce a spectacular effect in the lower rainfall areas, whereas a large increase is necessary before there will be any significant effect on the vegetation in the higher rainfall areas, except possibly for an extension of the riverine vegetation.

The calculations of Warren (1970) were based on his work in the sand dune areas of northern and central Sudan and are not necessarily applicable to southern Sudan. Unfortunately there are no data available regarding Quaternary climatic fluctuations in the south. Consequently the reconstruction of the vegetation has assumed a parallel shifting of the climatic and vegetation belts throughout the Sudan. This is probably unlikely for the maximum northerly shift of the climatic belts in the south was probably greater than the 250 km suggested by Warren for northern and central Sudan. However, without additional data, it seems preferable to follow Warren rather than to speculate further.

It has also been assumed that the changes in vegetation would at first lag behind the alterations in climate. With increasing aridity many deep-rooted species would tend to hang on, perhaps for several generations, until the underground water table could no longer be reached; conversely, with increasing rainfall, colonization would be delayed because of the time required for suitable habitats to develop and suitable seeds to become available. The vegetation maps consequently represent an arbitrary decision as to a period when the vegetation might have reached its maximum development. Since there are virtually no factual data available regarding the plant species present in the past, the vegetation can only be described in very general terms and compared with what are regarded as 'climax' communities today. There is fortunately considerably more fossil evidence regarding the fauna (Table 1), as well as historical information on the former distributions of some of the mammals, thus enabling a more realistic appreciation of the palaeo-fauna to be made.

3.3.2 *Present-day climatic shifts as a model for Quaternary palaeoclimatic changes*

Some of the assumptions and problems involved in modelling Sudanese vegetation changes in the manner attempted here have now been outlined. The rationale for such an attempt is the very close parallelism between vegetation zones and rainfall zones in the Sudan and indeed elsewhere in Africa. Experience during the recent 1968-1972 drought showed that parallel shifts in the rainfall belts took place throughout the Sudan and may be analogous to the Quaternary climatic changes with which we are concerned.

The reconstruction of the vegetation during past climatic periods is based on the assumption that the isohyets moved northwards or southwards parallel to their present-day positions. Such a system has the advantage of simplicity; there are insuperable barriers to attempting the reconstruction of large areas if a more flexible climatic regime is postulated.

Since the 1960's there has been a noticeable southward shift in the isohyets over northern tropical Africa by approximately 9 km per annum (Winstanley 1973). Consequently the comparison of these recent changes in the average rainfall, using north to south transects within narrow longitudinal limits, seems a plausible means of testing the possible validity of the hypothesis. Two

Table 3.2. Rainfall during and before the 1968-1972 drought in relation to latitude. Stations (A) give a north-south transect along the Nile valley; stations (B) a north-south transect from Kordofan southwards.

Station	Latitude	Longitude	Pre-1965 years	Average rainfall (mm) mm	1965-1972	1968-1972
A. Abu Hamed	19°32′	33°19′	30	19	8	8
Karima	18°33′	31°51′	15	34	39	29
Atbara	17°42′	33°58′	27	72	54	43
Shendi	16°42′	33°26′	14	136	69	56
Khartoum	15°36′	32°33′	46	163	157	128
Wad Medani	14°24′	33°29′	33	396	324	301
Ed Dueim	13°59′	32°20′	33	307	229	171
Kosti	13°10′	32°40′	8	487	353	345
Malakal	9°33′	31°39′	30	817	726	708
Juba	4°52′	31°36′	26	971	992	967
B. Bara	13°42′	30°22′		304	213	218
El Obeid	13°10′	30°14′	34	362	266	257
Dilling	12°03′	29°39′		662	510	536
Kadugli	11°01′	29°43′	25	743	678	612
Wau	7°42′	28°01′	33	1 097	1 052	990

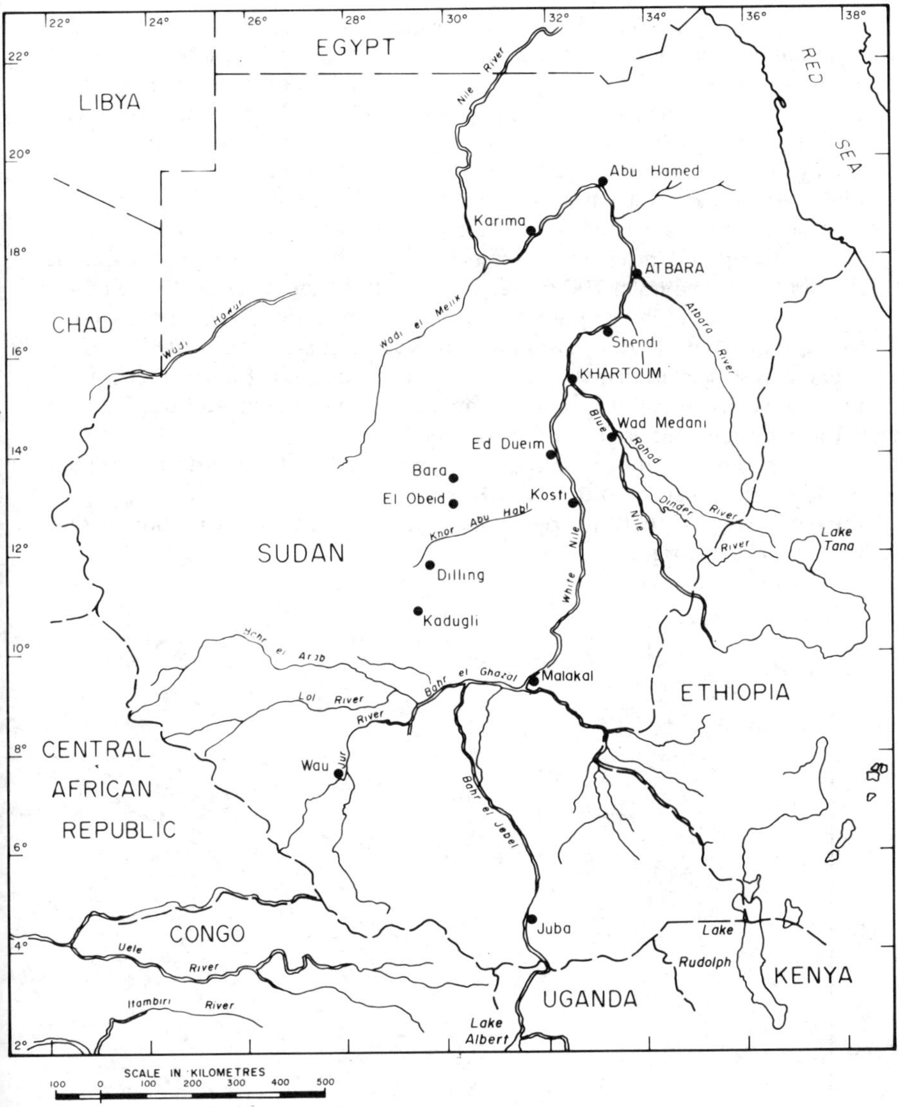

Figure 3.2. Map of selected rainfall stations in Sudan.

transects were therefore selected, one from Abu Hamed to Juba, along the Nile valley, the other from Bara to Wau, to the west of the Nile (Figure 2). The rainfall data are shown in Table 2; the pre-1965 figures were taken from Walter & Lieth (1967), the post-1965 figures were calculated from the US Department of Commerce Weather Bureau (1965-1972), supplemented by some unpublished data from the Sudan Meteorological Department.

Comparisons between the pre-1965 average rainfall and that for 1968-72 were obtained by plotting rainfall against the latitude of the meteorological stations to produce the graph shown in Figure 3. The overall impression is that the isohyets have moved southwards, parallel to their former positions.

The reason for the almost unchanged rainfall at Juba is not understood: it may be due to a purely local effect. Records from additional stations between Malakal and Juba and to the south of Juba are obviously required.

The relatively small change in the average rainfalls to the north of Shendi is due to unusually heavy cloudbursts in some years. For example, at Karima no rain fell in 1969, yet 100 mm fell in July 1970 and most of that fell during a few storms. In such low-rainfall areas it is the effectiveness of the rainfall that is important, not only for individual years but also over a number of years. It is suggested that the occasional freak storm, as at Karima, is unlikely to have much effect on the climatic deterioration of the vegetation, since it is the dry years that are the effective agent.

The method is crude; the pre-1965 average rainfall for Khartoum is based on 46 years of records, but that of Kosti on only eight years. Nevertheless, the results suggest that the original hypothesis for the reconstruction of the vegetation is justified.

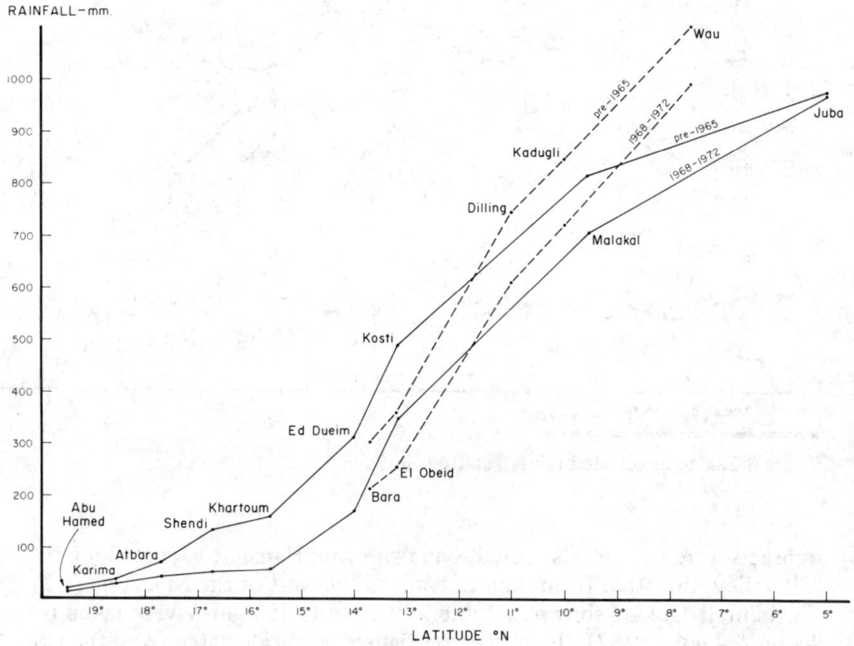

Figure 3.3. Rainfall changes during the 1968-1972 drought as a function of latitude, central Sudan.

3.3.3 *Present vegetation and the influence of man*

Chapter 8 discusses the present vegetation of central Sudan in some detail. The map of the major vegetation units of the Sudan (Figure 4) is reduced from that of Harrison & Jackson (1958). The present summary of relevant vegetation units in the northern and central Sudan is based upon Harrison & Jackson (1958), supplemented by unpublished reports by Hunting Technical Services and by Wickens' observations in Darfur and Kordofan.

3.3.3.1 *Desert.* In its extreme form, desert is regarded as being completely hostile to plant and animal life. Although an average rainfall of 75 mm is accepted as delimiting the desert region, the transition from desert to semi-desert is so gradual as to escape definition.

Occasional showers, perhaps after an interval of several years, can briefly transform areas of apparently barren desert into a valuable grazing area known as 'gizzu'. Any permanent vegetation is confined to the seasonal watercourses, which could perhaps be regarded as an intrusion of the semi-desert into the desert proper.

3.3.3.2 *Semi-desert scrub and grassland.* This consists of either grasses and herbs without any woody species or, more usually, with a variable scatter of low bushes up to 2 m high. The rainfall, which is between 75 and 400 mm, is too low for the herbage to be sufficiently dense for grass fires to become an annual possibility.

On sand dunes, both in the lower rainfall areas as well as higher rainfall areas where the sands are either unstable or excessively drained, *Panicum turgidum* is likely to be the dominant grass. *Panicum turgidum* is regarded as an excellent grass for stabilizing sand as well as being a good browse plant for ruminants throughout the year. The tussocks are usually well spaced, with plenty of bare ground between. It is often the sole perennial grass present, with a number of herbs and annual grasses thinly scattered between the tussocks. Woody associates may include *Leptadenia pyrotechnica* and *Acacia tortilis* subsp. *tortilis.*

On the stabilized sands in the higher rainfall areas, the perennial grass, *Aristida sieberana (A.pallida)* is likely to be dominant. The tussocks are usually sufficiently close for the spreading panicles to obscure the ground below. The grazing, although relatively abundant, is probably less succulent and less palatable than that of *Panicum turgidum.*

Woody associates may include *Leptadenia pyrotechnica, Acacia senegal* and *A.tortilis* subspp. *spirocarpa* and *raddiana.* With increasing rainfall, the woody species, which were previously confined to the seasonal watercourses, gradually became more widespread until there is no apparent demarcation between the semi-desert scrub and thorn savanna.

On clay soils the dominant grass is likely to be *Schoenefeldia gracilis,* an unpalatable annual, with *Sehima ischaemoides* locally dominant. The grasses tend to be restricted to areas where there is a slight accumulation of shallow drift sand. *Acacia mellifera,* which at first is confined to the seasonal water-

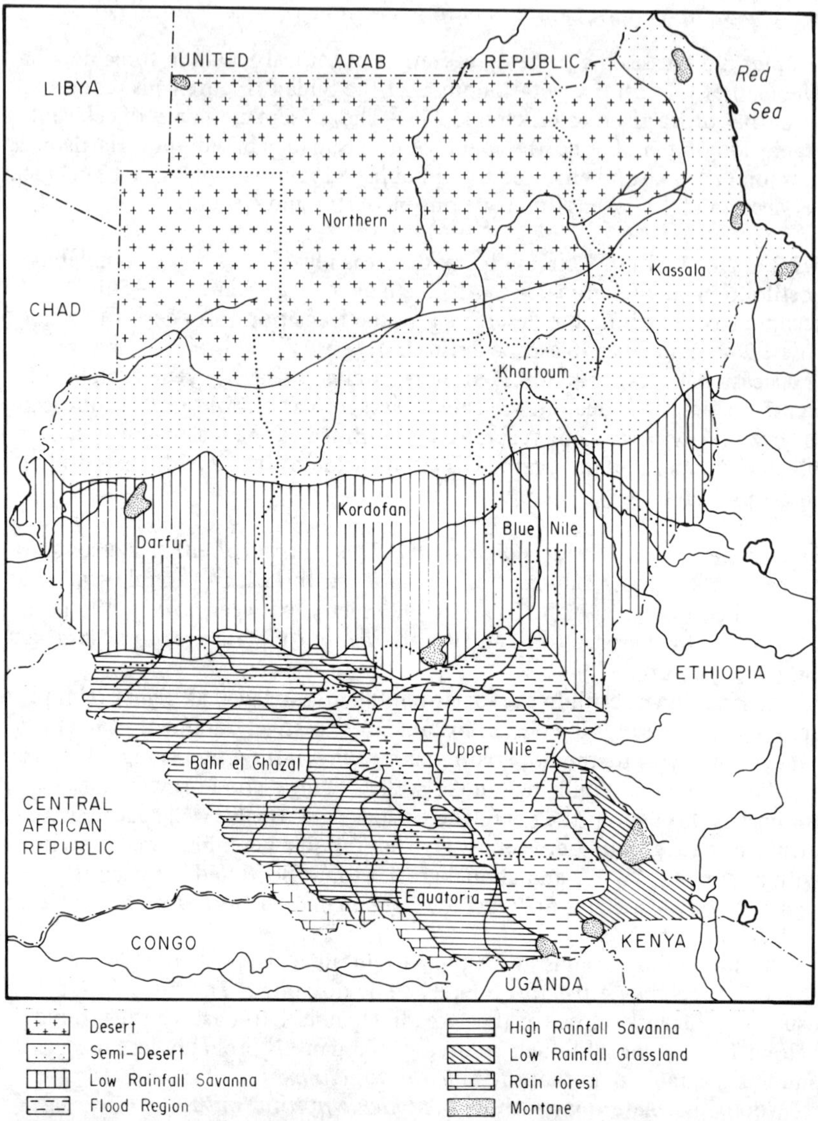

Figure 3.4. Present-day vegetation regions of the Sudan. (Based upon Harrison & Jackson 1958; unpublished data in Hunting Technical Services reports 1964-1966, and G.E.W.'s observations in western Sudan.)

ways, slowly becomes more widespread with increasing rainfall, until there is a gradual transition to thorn scrub at about the 400 mm isohyet. The clay soils, although inherently more fertile than the sands, are less productive due to the low availability of the soil moisture.

3.3.3.3 *Low rainfall savanna.* In the drier parts of this subdivision thorny trees and shrubs of *Acacia* predominate to form what is termed thorn savanna and scrub. In the wetter areas deciduous savanna woodland of predominantly broad-leaved deciduous trees occur.

3.3.3.4 *Thorn savanna and scrub.* The main woody constituents are species of *Acacia,* and the herbage is usually sufficiently dense for grass fires to be an annual possibility. On the sandy soils *Leptadenia pyrotechnica* and *Acacia tortilis* subsp. *raddiana* are likely to be the main woody constituents of the drier areas, with *Aristida sieberana* as the dominant grass. Further south the trees are replaced by *Acacia senegal* and *Combretum glutinosum (C.cordofanum).* Much of the dune area today is dominated by almost pure stands of *Acacia senegal,* the producer of high quality gum arabic. The stands are obviously the result of a long period under a gum acacia-cultivation cycle (Jackson & Shawki 1950), so that it is virtually impossible to determine what the climax vegetation may have been. On the clay soils the dominant shrub or small tree of the drier areas, with a rainfall of between 400 and 500 mm, is *Acacia mellifera,* which may form dense, pure stands of almost impenetrable thicket. *Schoenefeldia gracilis* and annual species of *Aristida* are the main grass constituents on the shallow accumulations of sand overlying the clays.

With increasing rainfall there is a gradual transition to dominance by *Acacia seyal.* At first *Acacia seyal* is confined to water-receiving sites, eventually, with increasing rainfall, forming pure stands away from the depressions; sometimes it is in association with the fire-resistant *Balanites aegyptiaca.* The grasses tend to occur in pure stands of *Hyparrhenia anthistirriodes* or *Cymbopogon nervatus,* with *Sorghum* spp. in the wetter areas. In the wetter areas the red-barked *Acacia seyal* var. *seyal* may be replaced by the white-barked var. *fistula,* also distinguished by the ant-galls at the base of the spines.

The fauna of the Dinder National Park (Cloudsley-Thompson 1973) would suggest that the *Acacia seyal* association has a high game carrying capacity. The historic record (Table 1) seems to suggest that the very large herds of buffalo, elephant, etc. were not so common on the sands to the west of the White Nile.

Today there are very few permanent lakes amongst the sand dunes of Kordofan, such as at Umm Badr (14°14′N, 17°57′E), and there is no major river system draining the sands. In the past there is good evidence that there were numerous shallow lakes between the dunes (Hunting Technical Services 1964a); some were obviously large, permanent lakes, as at Sodiri (14°25′N, 29°05′E) from which the subfossil mollusc fauna has been recorded by Huckriede & Venzlaff (1962).

3.3.3.5 *Deciduous savanna woodland.* On soils derived from the Nubian sandstone or basement complex with a rainfall of between 450 and 650 mm, *Combretum glutinosum, Terminalia brownii, T.laxiflora* and *Sclerocarya birrea* are among the more important trees. *Aristida sieberana* would probably be the dominant grass but for the annual grass fires that sweep through the area.

Between 650 and 800 mm, *Anogeissus leiocarpus* also occurs with the above-mentioned trees, and may even become locally dominant in areas receiving over 800 mm. The grasses include *Hyparrhenia* spp., *Andropogon gavanus* and *Ctenium elegans.*

On clay soils, with a rainfall of 800 mm and more, the dominant trees are *Combretum hartmannianum* and *Anogeissus leiocarpus,* usually with some *Acacia seyal,* which may be locally dominant. There are also sporadic occurrences of *Sterculia setigera.*

3.3.3.6 *Swamp and wetland savanna.* This vegetation type has been more thoroughly investigated than any other in the Sudan (see Chapter 8). It consists of the permanent swamps of the 'Sudd' as well as the surrounding areas which are subjected to conditions ranging from seasonal flooding to that of having a high water-table with rare to occasional inundation.

A direct comparison between the Pleistocene swamps of the Gezira and the present-day papyrus swamps of the 'Sudd' seems reasonable, since the former northern limits of the 'Sudd' were probably once contiguous with the swamps lining the White Nile.

3.3.3.7 *Riverine woodland.* This community is described in detail in Chapter 8, and is confined mainly to the Nile and its tributaries, and to the more important rivers which flow down from the Jebel Marra massif.

3.3.3.8 *Hill and montane vegetation.* The low inselbergs or jebels that occur scattered about the Sudan, because of their wide range of ecological niches and relatively good protection from fire, sometimes with a higher soil fertility or more readily available soil moisture, usually have a richer flora than the surrounding plain. An analysis of their floras can sometimes help in determining past distributions of species. The relict gallery forest flora on Jebel Marra has been particularly helpful in this respect (Wickens 1975).

3.4 HYPOTHETICAL LATE QUATERNARY VEGETATION CHANGES IN CENTRAL SUDAN

3.4.1 *Hypothetical vegetation during the cool, dry terminal Pleistocene (20 000 - 15 000 BP)*

The reconstruction of the probable vegetation in central Sudan during this arid period is shown in Figure 5a. Desert conditions are believed to have prevailed over much of the area, and it is possible that the White Nile may have ceased to flow into the main Nile (Williams & Adamson 1974, 1980). West of the White Nile wind-blown sands covered the gently undulating late-Cainozoic surface (Hunting Technical Services 1964a), extending southwards to about latitude 13°N. The sands probably buried all previous vegetation. Even at the southernmost limit of the sand, with an estimated rainfall of about 75 mm, it seems doubtful if the sands would have weathered sufficiently for plant growth until the start of the following humid period.

There is no firm evidence that the sand sheet extended east of the White Nile. If not, then the soils derived from the underlying Nubian sandstone may have produced some ephemeral herbs and grasses following infrequent rain showers. Unlike the sterile desert to the west, there would have been some grazing for such desert animals as the oryx, addax and dorcas gazelle.

If the present-day isohyets (Randell 1961) can be regarded as a model for the conformation of past climatic conditions, then the desert may not have extended quite so far south to the east of the White Nile.

To the west of the White Nile there would have been an abrupt change from the recent sterile desert sands to semi-desert grassland and scrub on the weathered soils derived from the basement complex. The vegetation, both to the east and west of the White Nile would probably be a mixture of both perennial and annual grasses and herbs, with perhaps some trees scattered along the main drainage lines.

The only fossils for this period have been recovered from the alluvial clays of the Blue Nile at Singa and Abu Hugar. These have been described by Arkell (1949) and Arkell et al. (1951), and consist of the skull of Singa Man, some artefacts and a number of mammals, the latter listed in Table 1. It is possible that a somewhat similar fauna to that now found in the Wadi Howar area (Table 1) could have also roamed the semi-deserts during this period.

3.4.2 *Hypothetical vegetation during the warm, wet early Holocene (c. 11 000 - 7 000 BP)*

A possible reconstruction of the vegetation during this humid period is shown in Figures 5b and 6. The high levels in both the White and Blue Niles, with overbank flooding of the Blue Nile and impeded flow of the White Nile, plus the distribution of subfossil mollusca in the Gezira, suggest the presence of extensive swamps and lakes between the two Niles (Williams & Adamson 1973, 1974; Adamson et al. 1974). Extensive papyrus swamps are believed to have extended along the White Nile as far as Khartoum. The Nile lechwe, which is today only found in the 'Sudd', has been found in the fossil assemblages of both Khartoum and Shabona (Table 1). Other mammals associated with such swampy habitats include the hippopotamus, cob, reed rat and mongoose, as well as the elephant and buffalo. The earlier and then reasonable interpretation of the 'forest form' of the buffalo, *Syncaros caffer aequinoctialis,* as indicating the presence of forest near Khartoum (Arkell 1949) does not agree with the present interpretation of the palaeoenvironment. This dilemma has now been satisfactorily explained by a recent revision of the African Buffalo (Grubb 1972) where *S.caffer aequinoctialis* is regarded as a savanna form. The distribution of snails in the Gezira area (Tothill 1946; Williams & Adamson 1973) is suggestive of temporary rather than permanent swamps, i.e. of 'toich' conditions.

The soils of the clay plain are liable to impeded drainage during the rainy season, consequently conditions would have been even more severe during the time of deposition. It is envisaged that conditions would have been somewhat similar to the Raqaba Catena of Harrison & Jackson (1958). This catena

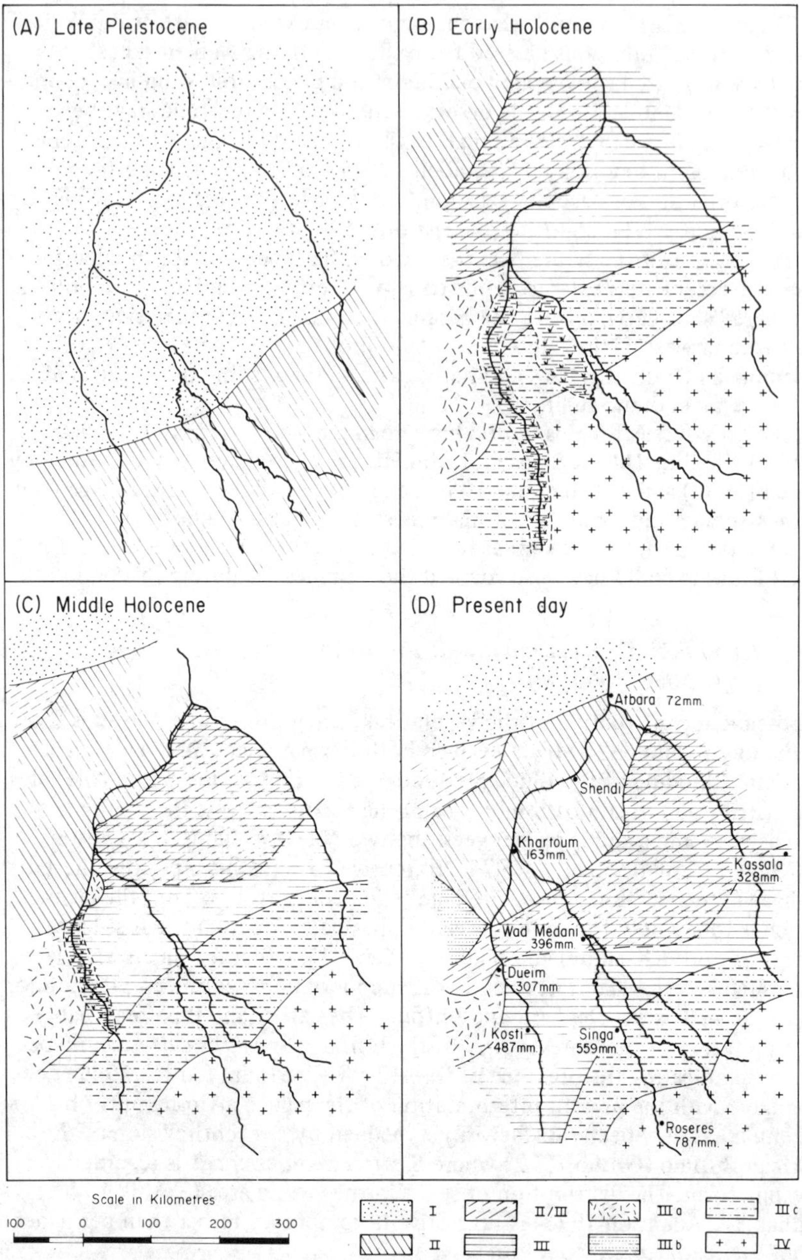

Figure 3.5. Hypothetical late Quaternary vegetation regions in central Sudan:
a) Late Pleistocene: vegetation belts shown 450 km south of their present position.
b) Early Holocene: vegetation belts shown 400 km north of their present position.

is the alluvial fan system of the Wadi Ibra from Jebel Marra draining into the Bahr el Arab and extensively flooded during the rainy season. It is envisaged that in the Gezira there may have been *Acacia seyal* alternating with open floodplain grassland in the north, while to the south there would be *Anogeissus leiocarpus, Combretum hartmannianum* and *Acacia seyal,* again alternating with floodplain grassland.

On the slightly higher ground both the dom palm, *Hyphaene thebaica* and the doleib palm, *Borassus aethiopum,* could have been present, although it is strange that there is as yet no evidence for their presence in the fossil record. Conditions may have been too wet for although both palms can tolerate a high watertable, they do seem to require some aeration, and are consequently intolerant of waterlogging.

On the high ground, woodland with *Anogeissus leiocarpus* may have occurred. Such sites would have been most suitable for permanent settlement, although the abundant insect life likely to be present in the surrounding marshy areas could have made life extremely unpleasant during the rainy season, if not for longer.

The distribution of the biting flies, the *Tabanidae,* is of considerable interest during the humid and sub-humid periods because they could have had a significant effect upon the distribution of man, his livestock and game, just as they do today. The following account is drawn almost entirely from Lewis (1952, 1953).

Although usually only troublesome during the rainy season, biting flies are present throughout the year along the White Nile between Shambe (7° 07′N, 30°46′E) and Jebelein (12° 36′N, 32°48′E); they are also more abundant in years of heavy rain. The northern limit for troublesome infestation approximates to the 500 mm isohyet. They probably attack most large mammals and, less readily, man.

The *Tabanidae* are a major source of irritation to animals. Their very presence can drive animals frantic. Their bite is extremely painful. These biting flies can also transmit such diseases as trypanosomiasis and anthrax. They

---

c) Middle Holocene: vegetation belts shown 250 km north of their present position.

d) Present-day: simplified reconstruction of climax vegetation using data from Harrison & Jackson 1958; Bunting & Lea 1962, and unpublished reports by Hunting Technical Services Ltd. 1963-1968.

Legend:

I.	Desert
II.	*Acacia tortilis – Maerua crassifolia* semi-desert scrub.
II/III.	Semi-desert grassland.
III.	*Acacia mellifera* thorn savanna.
IIIa.	*Acacia senegal* thorn savanna.
IIIb.	*Acacia mellifera – Commiphora* thorn savanna.
IIIc.	*Acacia seyal* thorn savanna.
IV.	*Anogeissus leiocarpus – Combretum hartmannianum* deciduous savanna woodland. Swamp symbols denote swamp vegetation along White Nile and 'toich' vegetation west of Blue Nile.

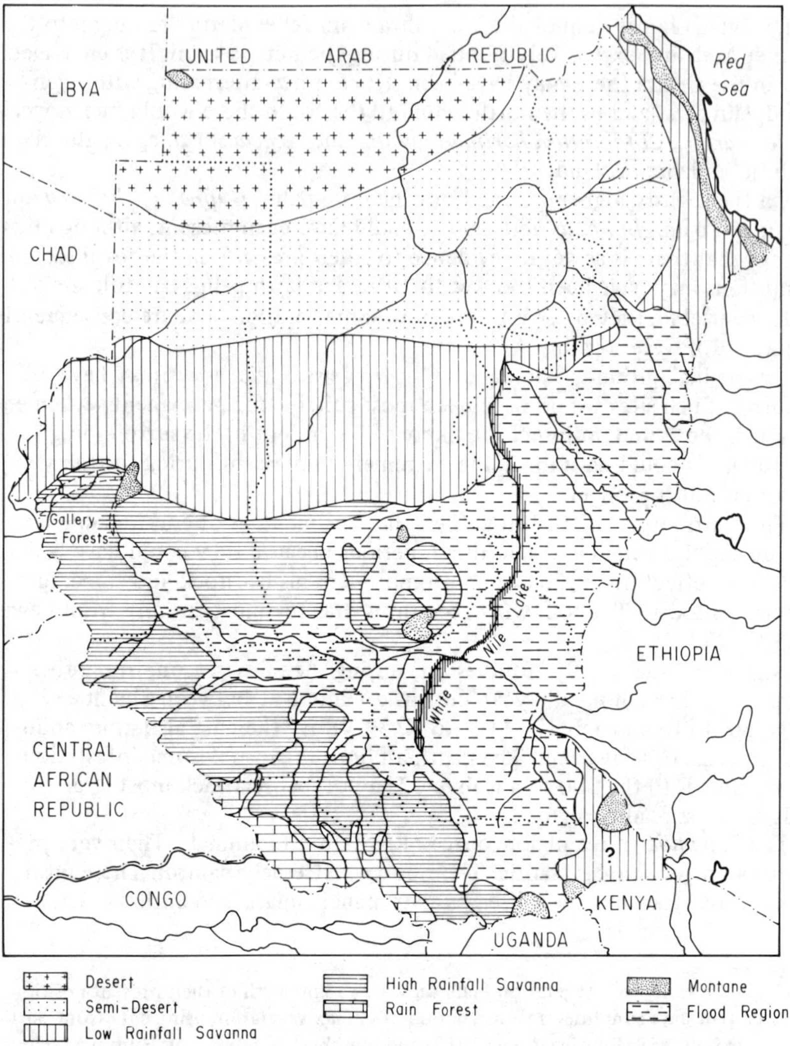

Figure 3.6. Hypothetical early Holocene vegetation regions of the Sudan (compare with Figure 3.4).

are believed to have been an important factor in limiting the Arab invasion and occupation of the Sudan. Even today the Arabs are forced to migrate with their cattle from the Bahr el Arab northwards during the rainy season. Because of the presence of the Arab tribes to the north, the Nilotic tribes are unable to migrate northwards. They move instead to higher ground. Even then they are only able to maintain their cattle by constant day and night attention, keeping the flies at bay with smoke. Although the Nilotic cattle tend to be more resistant to the *Tabanidae,* of all the domestic animals it is the goat that is the least affected.

In the past the presence of the *Tabanidae* could have been an important factor in agricultural development. During both the humid and sub-humid periods the 500 mm isohyet would have been in the vicinity of Khartoum; consequently the biting flies (and mosquitoes) were probably very troublesome throughout the Gezira area. Permanent settlement may have been possible on higher ground, such as around Jebel Moya, but some form of nomadic life seems more probable.

The domestication of cattle would certainly have been greatly retarded, if not impossible in the Gezira during this period. In fact, any herding of cattle would imply a good control of fire in order to keep up a constant supply of smoke. Of all the domesticated stock, preference would surely have been given to the goat, the animal that would be the most likely to survive. Evidence for the domestication of cattle should be sought either to the west of the White Nile or further to the north, in the areas where the biting fly is likely to be less troublesome.

The wild sorghums are usually to be found on the wetter clay soils; thus it seems likely that the development of cultivated strains could have been seriously delayed in the Gezira because of the physical discomfort of living in the area during the rainy season, when the wild sorghums would be growing.

To the west of the White Nile the raw, sterile sands should have weathered sufficiently to be stabilized by grasses and support thorn savanna, with *Acacia senegal* and *Combretum glutinosum* to the north of the 450 mm isohyet, and deciduous savanna woodland containing *Terminalia brownii* and *Anogeissus leiocarpus* to the south.

The dune sands are free draining, but between the dunes there are numerous clay depressions. With few exceptions, these are now dry. In the past they would have contained small lakes or swamps, as is evident from the presence of sub-fossil mollusca in the ancient lake depressions near Jebel Aulia (Williams et al. 1974) and at Sodiri (14°15′N, 27°05′E), noted by Huckriede & Venzlaff (1962), as well as the undated fragments of the rhizome of *Phragmites* sp. from Mazrub (Hunting Technical Services 1964a). Some of the still remaining lakes, such as Umm Badr (14°14′N, 27°57′E) are fringed with sunt, *Acacia nilotica* subsp. *nilotica,* and it is assumed that others were similarly fringed in the past.

Further south, in the vicinity of En Nahud, the larger seasonal lakes are also fringed with sunt and may contain ambatch, *Aeschynomene pfundii,* which is somewhat similar to *A.elaphroxylon* of the 'Sudd'. It appears to decline with the silting up and drying out of the lakes. Both species are used for making canoe-rafts, as they must have been similarly used in the past.

Figure 6 represents a hypothetical reconstruction of the early Holocene vegetation zones for the Sudan as a whole. The map was compiled by relating the 250 km northerly shift in climate to a parallel shift in the present-day vegetation shown in Figure 4, which was adapted from Harrison & Jackson (1958), after taking into account the distribution of soils in the Sudan.

The increase in the area of the Flood Region corresponds to the present extent of the soils of the clay plain. The area is probably exaggerated since deposition of alluvium continued after the early Holocene. Nor can it be assumed that 'sudd' and 'toich' conditions were prevalent throughout the entire Flood Region. It is exceedingly difficult to visualize what conditions must have been like, although it may be presumed that varying degrees of seasonal inundation were likely over much of the area, rather as in the Baggara/ Raqaba 'Repeating Pattern' found on active floodplains to the south of Nyala today.

Riparian woodland or forest may have bordered the major waterways, as in the upper reaches of the Bahr el Arab today. Extension of the rain forest as gallery forests was possibly even more extensive than that already suggested for the Jebel Marra area, along the Wadis Azum and Bulbul (Wickens 1971). At present there have been insufficient detailed ecological surveys of the central and southern Sudan for the gallery forests to be located precisely.

In the more humid past a number of tropical savanna species must have entered Egypt via the Nile valley, and there is good evidence from species distribution to show that this did indeed occur. However, there is no indication that the Nile valley may have been used in the reverse direction as a migration route for temperate species. It is extremely unlikely that this could have taken place even during the height of the very wet phase; lowland conditions would still have been too severe.

The Red Sea Hills and the Nuba Mountains were undoubtedly better afforested. The former must have been the migration route along which temperate species reached the mountains and uplands of Ethiopia and central Africa, as is evident from the distribution of a number of such species (Wickens 1971). Indeed, the climate diagrams for the Red Sea area today (Walter & Lieth 1967) are suggestive of a very dry Mediterranean-type climate, which under higher rainfall conditions would have readily facilitated the migration of temperate species from the Mediterranean area.

The Nuba Mountains, like the Red Sea Hills, have now been devastated by man's activities. However, there is an interesting northern outlier, Jebel Ed Dair, which so far appears to have escaped the full force of man's activities and should be worthy of investigation.

3.4.3 *Hypothetical vegetation during the mid-Holocene onset of desiccation (c. 5 000 - 3 000 BP)*

The vegetation for this sub-humid period is shown in Figure 5c. The mammal fauna at Jebel Tomat, and also of Esh Shaheinab, are mainly of savanna-inhabiting species (Table 1), suggesting that the swamps of the White Nile were probably much reduced in area. The mollusc fauna (see Chapter 9) is similar to that of Early Khartoum, and suggests seasonal rather than permanent swamps, and seasonal swamps are not the ideal habitat for the Nile lechwe and other swamp-loving animals. The presence of Nile lechwe in the fossil fauna of Khartoum is not a contradiction, for the permanent swamps of the White Nile were quite near; but the journey from the swamps to Esh Shaheinab would probably take a laden hunter two days, so that the absence of fossil Nile lechwe bones is to be expected.

From the data at present available, the significance of the Jebel Tomat fauna (Table 1) is not clear, but perhaps there was a drastic reduction in the extent of the permanent swamps.

Thorn savanna is believed to have been prevalent throughout much of the area, with *Acacia seyal* on the clay soils and *A.senegal* on the sands, which would also include the dunes between Jebel Aulia and Hashaba. The presence of Neolithic and older artifacts on these dunes (Williams 1966 and 1968) implies stabilization of the dunes and the presence of trees because man would certainly avoid unstable sands (the unstable dunes of the Kheiran to the west of Bara (13°42′N, 30°32′E) are certainly unoccupied today), and would not have a settlement, no matter how temporary, far from fuel supplies, especially in an area where the biting flies are so abundant.

The results of Professor J.D. Clark's archaeological investigations at Jebel Tomat indicate that the Neolithic inhabitants were hunters, fowlers and fishermen, cultivated sorghum, and owned cattle and sheep/goats (Clark 1973). The agricultural activities represent a late phase in man's activities, when the climate was drier, but not as dry as today.

In his role as hunter and fisher, man probably had very little effect upon the environment. He was mobile, there were no crops to guard, thus he could readily seek out fresh hunting grounds as game became scarce or migrated. The presence of beads of amazon-stone from Tibesti (Arkell 1961) is evidence of the extent of Neolithic man's wanderings: there was certainly no waterless desert then to the west of Khartoum to prevent his wanderings.

The acquisition of domestic stock would encourage the development of a seasonal migration similar to that of the Baggara Arabs of today, and for the very same reasons. Again it is doubtful if the herds were ever sufficiently large to cause any overgrazing problem. The numerous epizootic diseases would ensure that stock numbers would remain at a low level. Overgrazing did not become a serious problem in the Sudan until about the 1930's when prophylactic medicine was introduced by the veterinarians. Even the goat would have done little damage to the vegetation under the conditions then prevailing. The goat is really wrongly accused of despoiling vegetation. It is only when the chances of natural regeneration are greatly reduced, and that usually by the activities of man, his grazing animals and fire, that the goat is likely to become a menace. The goat is able to survive in areas where other domestic animals would starve, and it is this ability to survive by seeking out widely scattered new growth that is so destructive to regenerating vegetation (FAO 1953).

With the introduction of cultivation there must follow either permanent or semi-permanent settlements. The clearing of land for cultivation takes time and labour, and consequently it is uneconomical to abandon after taking only one crop. Depending upon the implements available, partial or total clearance of land must involve the destruction of vegetation to such an extent that at least a decade is necessary for its recovery to its 'original' state.

With settlement there is also a natural tendency for a number of domestic animals to be kept more or less permanently nearby, which in turn begins to create problems of over-grazing as well as delaying the natural regeneration of fallow land. A further effect of settlement is the demand for readily available supplies of timber for both fuel and construction purposes. This usually results in the wholesale destruction of all the trees in the immediate vicinity.

With the development of cultivation goes the necessity to protect the growing crops from the depredations of grazing or merely wandering animals. A single hippopotamus can ruin a growing crop overnight; consequently such animals are either destroyed or driven away. Livestock, too, has to be protected from predatory animals. Thus, there develops a change in the role of man in relation to the environment, from being in equilibrium to domination.

3.4.4 *Hypothetical vegetation during the late Holocene: the accelerated impact of man (3 000 BP to present)*

As the climate became progressively drier, the thorn savanna belt retreated southwards, eventually to be replaced by the semi-desert scrub and grassland of today (Figure 5d). The change in the climate was a gradual process, passing almost unnoticed. It is therefore likely that man would have remained in his original settlements and carried on his former activities against an increasingly hostile environment until eventually forced to move by shortage of water and food. In ecologically marginal areas, man's struggle to survive usually results in accelerated erosion, thereby further aggravating the situation. An additional complication is that occupation of land, whether by a family or a tribe, involves territorial rights, so that any migration of man consequent upon an increasingly hostile environment is likely to create social problems.

Probably the two most significant events in the biological history of the Sudan have been the introduction of firearms during the 17th century and the widespread use of preventative medicine by man and his domestic animals during the present century. The first has led to an increasing ease by which man has been able to destroy game animals, the second to overpopulation by man and stock.

Observations by Kassas (1968) show that the vegetation boundaries established by Harrison & Jackson (1958) have shifted southwards during the past decade. 'This means that species that formerly abounded in lower rainfall areas failed to regenerate in their own country and moved towards areas with higher rainfall. This is a replacement process with a general effect of shift of vegetational zones. This is apparently caused by combination of several processes including differences in tolerance of destructive agencies (fire, cutting, overgrazing, etc.) and differences in regeneration capacity under harassing influences of human interference.'

3.5 SUMMARY AND CONCLUSIONS

In the preceding sections of this chapter we have reviewed a wide array of scattered evidence relating to the past and present distribution of the fauna and flora of central Sudan. We will now summarize some of the more important conclusions.

1. During the cool, dry and windy terminal Pleistocene (c. 20 000 to c. 12 000 BP) much of central Sudan was arid and much of south-central Sudan was probably semi-arid. Rainfall was lower and more seasonal than now. If we accept that the southern limit of active wind-blown dune-sands in Kordofan and Darfur coincided with the 150-200 mm isohyet, a *southerly* shift of the vegetation belts of *up to 450 km* beyond present-day limits seems plausible.

2. The early Holocene period (c. 11 000 to c. 7 000 BP) was warm and wet in central Sudan. Rainfall was both more effective and absolutely higher than now. Grasses colonized and stabilized previously active dunes. Small lakes and swamps were common in the swales between the dunes. A northerly shift of vegetation belts of 250 km beyond present limits seems reasonable for the wettest part of this period. The mesophytic forest elements which now occupy deep, secluded valleys along the slopes of Jebel Marra may well date from this phase.

3. By late-middle Holocene times (c. 4 000 to 2 000 BP) climatic desiccation had set in. Lakes and swamps dried out west of the White Nile. Although still somewhat wetter than now, the vegetation belts were retreating southwards. Pastoralism and agriculture, hitherto made difficult by the presence of biting flies and associated diseases, became more common throughout the central Sudan. The influence of man upon the fauna and flora became increasingly pronounced, particularly since his delayed occupation and cultivation of the Gezira coincided with progressive deterioration of the climate.

4. The hypothetical shifts in vegetation regions assume that parallel shifts in the vegetation are related to parallel shifts in the rainfall belts. The behaviour of the rainfall zones during the 1968-1972 drought offers a modern analogue to what may have happened in the past. The reconstructions take into account the possible modifying influence of soil differences, topography, and man. For example, an acacia growing on sand requires only two-thirds of the rain needed by a similar species of acacia growing on clay. The modern 'climax' vegetation is much disturbed by irrigation in the Gezira and by grazing and cultivation elsewhere in central Sudan, but there is still a very definite correlation between climate and vegetation in the Sudan as a whole.

5. Although there is an ever increasing number of archaeological and geomorphic studies being carried out in the Sudan, there is still very little information available regarding the palaeobiology. The faunal records have been summarized by Whiteman (1971); the palaeoflora remains virtually unknown. There is an urgent need for fossil pollen studies to be made throughout the Sudan, to confirm or destroy the hypothetical reconstructions now being made.

3.6 REFERENCES

Adamson, D., J.D. Clark & M.A.J. Williams 1974, Barbed bone points from Central Sudan and the age of the 'Early Khartoum' tradition. *Nature* 249:120-123.

Arkell, A.J. 1948, The historical background of Sudan agriculture. In J.D. Tothill (ed.), *Agriculture in the Sudan.* Oxford University Press:9-17.

Arkell, A.J. 1949, *Early Khartoum.* Oxford University Press.

Arkell, A.J. 1953, *Shaheinab.* Oxford University Press.
Arkell, A.J. 1961, *A history of the Sudan from the earliest times to 1821.* London, Athlone Press.
Arkell, A.J., D.M.A. Bate, L.H. Wells & A.D. Lacaille 1951, The Pleistocene fauna of two Blue Nile sites. *The fossil mammals of Africa* 2. London, British Museum (Natural History).
Bagnold, R.A. 1954, The physical aspects of dry deserts. In J.L. Cloudsley-Thompson (ed.), *Biology of deserts.* London, Inst. Biology:7-12.
Baker, S.W. 1867, *The Nile tributaries of Abyssinia.* London, Macmillan.
Bate, D.M.A. 1951, The mammals from Singa and Abu Hugar. *The fossil mammals of Africa.* 2. London, British Museum (National History):1-21.
Birks, H.J.B. & R.G. West 1972, *Quaternary plant ecology.* 14th Symposium of the British Ecological Society, Univ. of Cambridge, 28-30 March 1972. Blackwell Scientific Publication.
Bruce, J. 1790, *Travels to discover the source of the Nile.* Edinburgh.
Bunting, A.H. & J.D. Lea 1962, The soils and vegetation of the Fung, east central Sudan. *J. Ecol.* 50:529-558.
Clark, J.D. 1973, Preliminary report of an archaeological and geomorphological survey in the central Sudan, January to March, 1973. Berkeley, mimeographed.
Cloudsley-Thompson, J.L. 1966, The fauna of the Merkhiyat Jebels. *Turox News* 44:26-30.
Cloudsley-Thompson, J.L. 1967, *Animal Twilight.* London, Foulis.
Cloudsley-Thompson, J.L. 1968, The Merkhiyat jebels. A desert community. In G.W. Brown, jr. (ed.), *Desert biology.* vol. 1. New York, Academic Press.
Cloudsley-Thompson, J.L. 1973, Developments in the Sudan Parks. *Oryx* 12:49-52.
Ensor, F.S. 1881, *Incidents on a journey through Nubia to Darfoor.* London.
FAO 1953, Grazing and forest economy, FAO Forestry and Forest Products Studies 4.
Friederichsen, L. 1878, Dr J. Pfund's Reisebriefe aus Kordofan und Dar-Fur 1875-76. *Mittheil. Geogr. Ges. Hamburg 1876-77:*121-305.
Grabham, C.W. 1920. Desiccation. *Sudan Notes and Records* 3:130-131.
Grove, A.T. & A. Warren 1968, Quaternary landforms and climate on the south side of the Sahara. *Geogr. J.* 134:194-208.
Grubb, P. 1972, Variation and incipient variation in the African Buffalo. *Z. Säugetierk. Berlin* 37:121-199.
Harrison, M.N. & J.D. Jackson 1958, Ecological classification of the vegetation of the Sudan. *Forests Bull.* 2 (new series), Republ. Sudan, Min. of Agric., Forests Dept.
Huckriede, R. & H. Venzlaff 1962, Über eine pluvialzeitlich Molluskenfauna aus Kordofan (Sudan). *Paläontol. Z.* (H. Schmidt Festband):93-109.
Hunting Technical Services 1964a, Land and water use survey in Kordofan Province of the Republic of the Sudan. Report on the geology, geomorphology and soils, vegetation and present land use. Doxiades Assoc. Athens. Doc. Dox-Sud-A26.
Hunting Technical Services 1964b, Roseires Soil Survey. Report 6. The White Nile, East Bank, Rabak to Khartoum. Soils and engineering reconnaissance.
Hunting Technical Services 1965a, Report 11. White Nile, East Bank, Melut to Rabak. Exploratory soils and engineering survey.
Hunting Technical Services 1965b, Report 14. White Nile, East Bank, Rabak to Khartoum. Semi-detailed soil survey and land classification.
Hunting Technical Services 1966a, Report 12. Blue Nile, East Bank, Roseires to Sennar, West Bank, Abu Hugar to Sennar. Semi-detailed soil survey and land classification; vol. 1, East Bank, Roseires to Sennar.
Hunting Technical Services 1966b, Report 13. Guneid to Hawata extension to the Rahad Project. Soil survey and land classification, vol. 1. The Report.

Idrisi 1836, *Géographie d'Edrisi,* vol. 1, (transl. P.A. Jaubert). Paris.
Jackson, J.K. 1957, Changes in the climate and vegetation of the Sudan. *Sudan Notes and Records.* 38:47-66.
Jackson, J.K. & M.K. Shawki 1950, Shifting cultivation in the Sudan. *Sudan Forests Dept. Mem.* 2. Khartoum.
Kassas, M. 1968, Dynamics of desert vegetation; Proc. Tech. Meeting on Conservation and Ecology, West Mediterranean, North Africa, Hammamet, Tunisia, March 1968 (mimeo).
Kirwan, L.P. 1957, Rome beyond the southern Egyptian frontier. *Geogr. J.* 123:13-19.
Lewis, D.J. 1952. Early travellers' account of surret flies *(Tabanidae)* in the Anglo-Egyptian Sudan. *Sudan Notes and Records* 33:276-298.
Lewis, D.J. 1953, *Tabanidae* of the Anglo-Egyptian Sudan. *Bull. Ent. Res.* 44:175-216.
Monod, T. 1964. The late Tertiary and Pleistocene in the Sahara. In F. Clark Howell & F. Bourlière (eds.), *African ecology and human evolution.* Methuen:117-229
Petherick, J. & Katerina Petherick 1869, *Travels in central Africa and explorations of the Western Nile tributaries.* 2 vols. London, Tinsley Bros.
Randell, J.R. 1961, A rainfall map of the Sudan Gezira. *Sudan Notes and Records* 42: 29-36.
Schweinfurth, G. 1873, *The heart of Africa. Travels and adventures from 1868 to 1871.* 2 vols, (transl. E.E. Frewer). London, Sampson Low & Co.
Shaw, W.B.K. 1936, An expedition in the southern Libyan desert. *Geogr. J.* 87:193-221.
Smith, J. 1949, Distribution of tree species in the Sudan in relation to rainfall and soil texture. *Sudan Govt Min. Agric. Bull.* 4.
Tothill, J.D. 1946, The origin of the Sudan Gezira clay plain. *Sudan Notes and Records.* 27:153-183.
US Department of Commerce Weather Bureau 1965-1972, Monthly climatic data for the world. Vols. 18-25.
Walter, H. & H. Lieth 1967, *Klimadiagramm-Weltatlas.* Jena, Fischer.
Warren, A. 1970, Dune trends and their implications in the central Sudan. *Z. Geomorph. NF, Suppl.* 10:154-180.
Werne, F. 1849, *Expedition to discover the sources of the White Nile in the years 1840, 1841.* (transl. C.W. O'Reilley). London.
Whiteman, A.J. 1971, *The geology of the Sudan Republic.* Oxford, Clarendon Press.
Whyte, R.O. 1951, Management and conservation in the Anglo-Egyptian Sudan. *Commonw. Bureau of Pastures and Field Crops Bull.* 41:69-79.
Wickens, G.E. 1970, J.D.C. Pfund, a botanist in the Sudan with the Egyptian military expeditions 1875-76. *Kew Bull.* 24:191-216.
Wickens, G.E. 1971, The flora of Jebel Marra (Sudan Republic) and its geographical affinities. Ph.D. thesis, Reading Univ.
Wickens, G.E. 1975, Changes in the climate and vegetation of the Sudan since 20 000 BP. *Boissiera* 24:43-65.
Wickens, G.E. 1976, Speculations on long distance dispersal and the flora of Jebel Marra, Sudan Republic. *Kew Bull.* 31:105-150.
Williams, M.A.J. 1966, Age of alluvial clays in the western Gezira, Republic of the Sudan. *Nature* 211:270-271.
Williams, M.A.J. 1968, A dune catena on the clay plains of the west central Gezira, Republic of the Sudan. *J. Soil Sci.* 19:367-378.
Williams, M.A.J. & D.A. Adamson 1973, The physiography of the Central Sudan. *Geogr. J.* 139:498-508.
Williams, M.A.J. & D.A. Adamson 1974, Late Pleistocene desiccation along the White Nile. *Nature* 248:584-586.
Williams, M.A.J. & D.A. Adamson 1976, The origins of the soils between the Blue and

White Nile rivers, central Sudan, with some agricultural and climatological implications. Econ. and Soc. Res. Council, Natl. Council for Res., Occasional Paper 6. Khartoum.

Williams, M.A.J., A.H. Medani, J.A. Talent & Ruth Mawson 1974, A note on the Upper Quaternary sub-fossil mollusca west of Jebel Aulia. *Sudan Notes and Records* 54: 168-172.

Winstanley, D. 1973, Rainfall patterns and general atmospheric circulation. *Nature* 245: 190-194.

J.R.VAIL
Portsmouth Polytechnic, UK

4

Geology of the central Sudan

4.1 INTRODUCTION

Sudan can be divided geologically into a number of clearly defined units. The central part of the country, that enclosed by the two Niles, contains elements of almost all the units known elsewhere.

In the relationship of geology to man's settlement the younger, unconsolidated units of river alluvium and superficial soils dominate, but man has made use of a variety of stones for implements, most of which are to be found in or close to the area under discussion.

The region reviewed here is centred on the Gezira plain, the land between the two rivers, and includes parts of the west and east banks of the White and Blue Niles respectively, from the vicinity of Kosti and Singa to Khartoum (Figure 1). At Khartoum the two rivers join to form the main Nile stream which flows unimpeded across a flat, broad valley for about 60 km and then it crosses over an inlier of older, crystalline rocks and through the centre of an ancient igneous complex, a barrier forming the Sixth Cataract at Sabaloka.

Rock outcrops over most of the area are not abundant, the country being largely covered by soils or sands and gravels. Across the northern sector arenaceous sediments of the Nubian Sandstone Formation crop out. These generally flat-lying strata in turn directly overlie the ancient crystalline metamorphosed and intruded basement complex rocks. North of Khartoum the latter are exposed in the Sabaloka inlier, but they also crop out south and east of Jebel Qeili in the Butana plains to the east of the capital. Scattered outcrops of basement gneisses are also to be found west of Ed Dueim along the White Nile, and these rocks also protrude through a younger cover of soils and silts in a belt between Kosti and Sennar in the Gezira plain, south of Manaqil.

Following the deposition of the Nubian Sandstone Formation there occurred a period of volcanic activity, represented here in the form of small basaltic plugs and sills, exposed west of Omdurman and at Sennar.

These events were followed by widespread erosion and subsequent deposition of silts and clays which, together with more recent river alluvium, wind blown sands, and residual gravels, make up a persistent cover several hundreds of metres thick in places, of superficial deposits which effectively obscure all older formations.

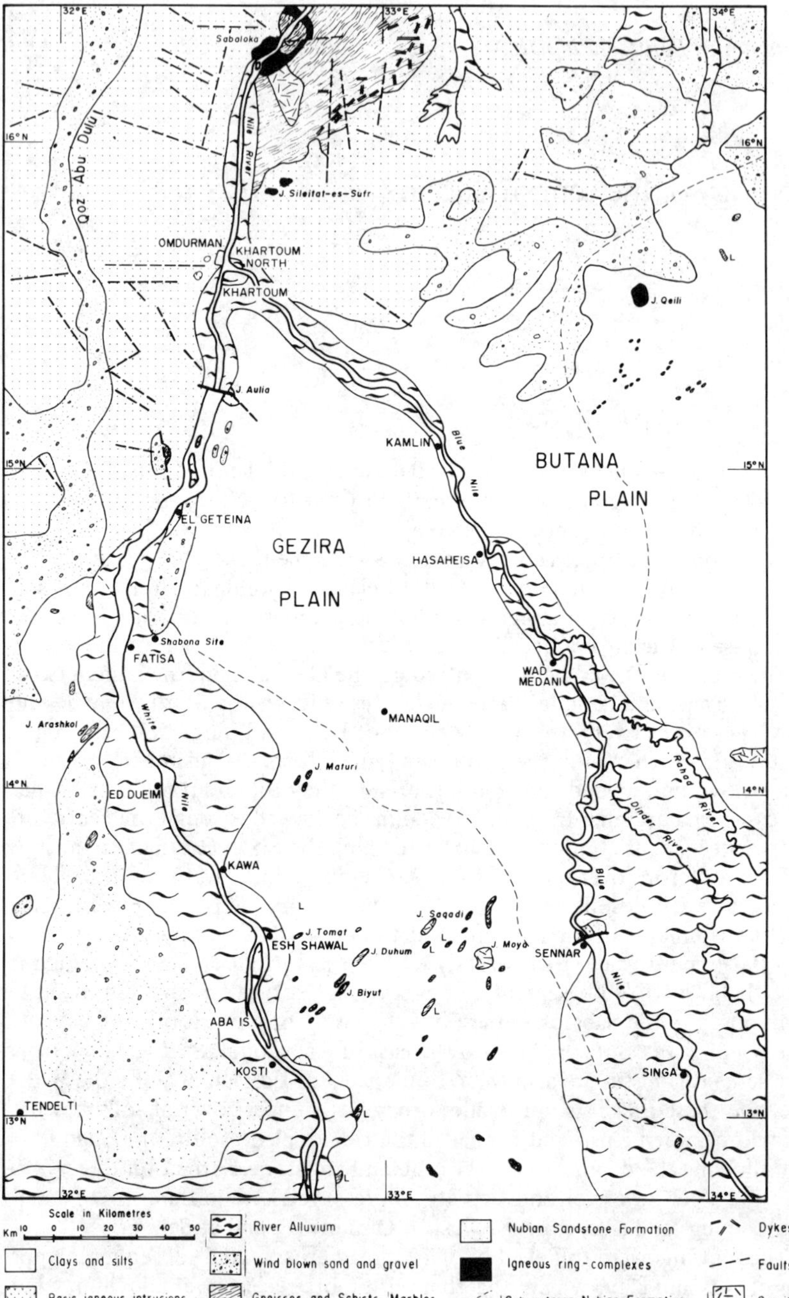

Figure 4.1. Geological map of the Gezira region of central Sudan (compiled by J.R. Vail 1973).

4.2 PREVIOUS GEOLOGICAL WORK

Geological mapping of Sudan was started about 70 years ago when a survey department was established. However, despite occasional mapping, especially for such projects as the construction of the Jebel Aulia and Sennar dams, and visits by prospectors, geologists, and archaeologists it was not until the 1950's that any systematic work was done in this part of the country.

The first geological map of the area was prepared by Delany (1952) when the Geological Survey Department published Sheet 55 – Khartoum, scale 1:1 million. On this Delany outlined the distribution of the major rock units. In 1953 Delany compiled for the Geological Survey a map of East Khartoum Province, scale 1:250 000, which included part of the area under review here and also showed sites of archaeological and prehistoric interest.

Since that time the Geological Survey has remapped part of the area, as yet unpublished, and a number of individual studies have been made, including investigations of scattered igneous localities, notably by Almond (1967, 1968), Ahmed (1968), Dawoud (1970), Kabesh (1967); underground water studies by Abdel Salam (1966) and El Boushi (1972); soil surveys by Hunting Technical Services (1964) and river studies by Berry & Whiteman (1968) and Omer El Badri (1972).

Despite these activities and also drilling for water the geology of the area remains poorly known, and little detailed study of petrography, structure or stratigraphy has been undertaken.

The present paper must be regarded as a general review only, as not all the area was visited by the author. However, it has been possible to examine a large number of aerial photographs, particularly US Air Force trimetrogons (1944-46), as well as Gemini, Apollo, and ERTS space imagery (1966-1973).

It is intended here to provide a geological background to the ecological and geomorphological studies, and it will be seen that most of the rock types have been used at some time in the past in the service of man.

4.3 OUTLINE OF THE GEOLOGY

The geology of central Sudan can be divided into the six main units first outlined by Andrew (1948) for the country as a whole. They are as follows:
F. Wind blown sand and superficial gravels.
E. Silts and clays of the Gezira plain and Umm Ruwaba Formation type.
D. Tertiary volcanic rocks.
C. Nubian Sandstone Formation sedimentary strata.
B. Post-orogenic igneous complexes.
A. Basement complex metamorphosed and igneous rocks.
Each unit is easily recognizable in the area (Figure 1) but the amount of outcrop and type of exposure varies considerably.

4.3.1 *Basement complex*

The oldest rocks exposed are an ancient crystalline group of gneisses and schists, metavolcanic rocks and granites.

Representatives of this unit crop out in four places, namely, in the Sabaloka inlier in the north, west of the White Nile between Omdurman and Ed Dueim, east of the Blue Nile in parts of the Butana plains, and south of the Gezira between Sennar and Kosti (Figure 1). Elsewhere they are covered by younger deposits, but have been encountered in depth in boreholes and wells throughout the area.

The Basement Complex group remains for the most part undifferentiated as insufficient work has been done in Sudan to distinguish recognizable groups within it. However, a number of distinctive rock types are to be found, and it may be that they will in future be recognized as belonging to specific periods of activity.

The main rock types of the basement in the Gezira region are as follows:

(i) *Gneisses and migmatites.* Much of the Sabaloka inlier is made up of a variety of quartzo-feldspathic gneisses, with some migmatites (Delany 1958; Dawoud 1970).

They form a monotonous succession which can be subdivided only with detailed mapping. The rocks are poorly to moderately foliated, generally grey in colour, medium-grained, and consist of quartz, orthoclase, muscovite and minor accessories like biotite, garnet and magnetite. They usually form massive, bouldery outcrops or are eroded to flat, sand covered plains.

Another type of gneiss occurs in the southern Gezira and SW of Fatisa at Jebel Arashkol (Figure 1). This is an augen gneiss of distinctive appearance. This rock is quarried from an isolated hill SE of Jebel Moya and used as railway ballast, consequently it has now been spread far across the area. It was also worked in ancient times, as several large grinding bowls have been found at Jebel Moya made of this rock type. It also occurs at Jebel Tomat, at which place it makes up part of the outcrop of the ridge.

It is a porphyroblastic quartzo-feldspathic gneiss of fine-grained dark groundmass and large feldspar augen. Under the microscope the rock is composed of large irregular, angular fragments of quartz, K-feldspar, altered plagioclase and perthite. The matrix is a granular mass of broken grains with sherds of brown-green mica, sericite, chlorite and accessory apatite, euhedral sphene, and streaky magnetite. The cataclastic texture is dominant and the rock grades from flaser gneiss to mylonite. It presumably represents a recrystallized cataclastic rock, formed from gneisses of the basement, which is now more resistant to erosion and hence protrudes above the general level of the plain.

(ii) *Granulites and charnockites.* East of Sabaloka, in apparent igneous contact with the gneisses, are a number of basic metamorphosed rocks. They are pyroxene-amphibole-garnet granulites, which have undergone metamorphism from nearby granitic intrusive rocks.

The only other place where metamorphic rocks of similar high grade are to be found is at Jebel Moya, 30 km west of Sennar. The mountain mass rises high above the flat clay plains. It is made up of bouldery outcrops of coarse-grained dark, granulitic rocks of typical charnockitic aspect. The charnockites are well-developed around the old Wellcome archaeological excavation site, and the house of boulders is made up almost exclusively of these rocks.

The rocks are composed of large crystals of hypersthene, plagioclase and quartz. In section pink-green pleochroic orthopyroxene, green amphibole and yellow-green biotite are surrounded by quartz and plagioclase. The latter is zoned, and has thin lamellae showing low extinction angles. Apatite and magnetite are also abundant.

The charnockites have been mentioned by Howie (1958) and by Kabesh (1967) and are of importance because of their high metamorphic grade. The nearest other examples of such rocks in the Sudan are to be found near Pibor Post in Equatoria, and near Wau in Bahr El Ghazal Province (Kabesh 1967).

(iii) *Granites.* A number of granitic masses have been distinguished in the Sabaloka basement (Dawoud 1970). They are now slightly foliated, massive, medium-grained grey or pink rocks.

Another band of granite, not visited by the author, occurs at Jebel Fau in the western Butana plain east of the River Rahad.

Granites are also important rocks at Jebel Moya, where they are intrusive into the charnockites, and also at Jebel Saqadi, where they make up the main ridge. In the past they were quarried here for rail ballast and for construction works at Sennar. The Jebel Saqadi granite is a distinctive rock. It is very coarse-grained, almost pegmatitic, porphyritic, orange-pink in colour and made up dominantly of quartz and feldspar. Under the microscope large, equigranular crystals of quartz, perthite, orthoclase and albite-oligoclase are associated with large, brown biotite crystals and minor dark green hornblende.

(iv) *Mica schists.* Two main types of mica schist occur in the area, and both are confined to the southern Gezira region, although similar rocks are well known from the Nuba Mountains of southern Kordofan, parts of the Butana plains, and along the Nile valley north of Atbara in Northern Province.

The rock types are sericite schists and graphitic schists. The former are light coloured, fine-grained rocks made up largely of quartz and white mica in variable proportions. They are very well foliated, and being soft do not form conspicuous outcrops. They evidently represent some original sedimentary rocks which have undergone a moderate grade of regional metamorphism.

The graphitic schists are dull, black rocks, also soft and of inconspicuous outcrop. They form a band between Jebel Moya and Jebel Saqadi.

(v) *Marbles.* At scattered localities in the Butana, near Jebel Qeili for instance, and in the southern Gezira several distinctive thin bands of white crystalline limestone occur. These rocks form an important source of lime for the cement factory at Rabak, opposite Kosti, the raw material being quarried further south on the east bank of the White Nile.

Small quarries were also opened about half way between Kosti and Sennar. Similar rocks occur near Atbara, far to the north, where they provide the material for the Portland cement works.

The rocks are generally white, cream or grey in colour and are fine- to medium-grained, crystalline, usually monomineralic calcite. Some dolomite is present in places, and with increase of impurities such minerals as tremolite, forsterite, diopside, and brown garnet may be present.

The marbles are in a moderate grade of regional metamorphism, and crop out in the vicinity of the graphitic schists. They are important for their source of lime, but there is no evidence that they were used at all by the ancients, no artifacts having been reported of this material.

(vi) *Greenschists.* A distinctive and important rock type from the Basement complex occurs in some of the archaeological sites. This is a dark greenstone of greenschist metamorphic facies. It is well known in parts of the Butana, especially to the east of the map area, and also is well developed in the Red Sea Hills, and north of Berber in Northern Province. It has not been identified *in situ* in the Gezira region.

The rock is a fine-grained metavolcanic of basaltic or andesitic aspect. Usually traces of original texture and structure have been obliterated by the light regional metamorphism that has affected these rocks, and the distinctive nature, low thermal grade, and lack of intense foliation suggest that this rock represents a younger, less metamorphosed group of basement rocks which probably overlie the more intensely tectonized gneisses, schists and granites.

The greenschists usually contain epidote, chlorite, altered hornblende and plagioclase laths. Some specimens are coarser grained and suggest altered dolerites.

They are particularly common at the Shabona site near Fatisa, which might be because a band of these rocks strikes south-westwards from the vicinity of Jebel Qeili beneath the Gezira plain between Kamlin and Hasaheisa towards Fatisa. They appear to be absent in the vicinity of Jebel Moya (Figure 1).

(vii) *Quartz veins and pegmatites.* The basement complex of Sudan is characterized by numerous small veins and bodies of quartz or quartz and feldspar, with occasional micas and other minerals. The Sabaloka inlier and southern Gezira are no exceptions. In Northern Province, near Shereik, the bodies contain concentrates of useful minerals like muscovite, beryl, tourmaline and apatite to make them of potential economic interest.

Several of the archaeological sites in the central Sudan, and much of the younger sedimentary formations have abundant fragments and pebbles of quartz. This quartz must have been derived from the basement rocks, and has in some instances been recycled to produce quartz gravels and sands which are now exploited for the building industry.

In the distant past they attracted little attention but provided suitable hard pebbles from which flakes could be chipped.

4.3.2 *Ancient post-orogenic central igneous complexes*

These igneous centres which post-date the metamorphism of the basement occur within the area of the map, and provide suitable material for prehistoric cultural use.

The centres are the large sub-volcanic complex astride the Nile at Sabaloka giving rise to the Sixth Cataract, the smaller buried complex of Sileitat-es-Sufur north of Khartoum North, and the ring-complex of Jebel Qeili in the western Butana plain.

The three complexes are possibly of similar age, although they are geologically somewhat different. The Sabaloka Complex (Delany 1958; Almond 1967, 1971; Sadig 1969) consists of a ring-dyke of porphyritic microgranite encircling a downfaulted central cauldron in which are preserved in the south a thick pile of associated extrusive rocks, and in the north the floor rocks of the gneissic basement. The volcanic pile is made up of a thin basal trachybasalt, overlain by fine, hard pink rhyolites, ignimbrites and quartz-feldspar porphyries. The latter acid rocks provide hard, brittle fine-grained material suitable for stone artifacts, pieces of which have been found in the Gezira. No obsidian has been located, the great geological age of the complex probably precluding preservation of such material.

Associated with the Sabaloka sub-volcano are a number of felsite dykes, intrusive into the crystalline basement rocks of the Sabaloka inlier to the east. They are generally fine-grained, pink rocks (Dawoud 1970) which have also been used to fashion axe heads and scrapers.

The Sileitat-es-Sufur complex is largely covered by Nubian Sandstone Formation sediments. It is of importance today for the quarry which supplies building aggregate and dressed stone for Khartoum. Much of the Jebel Aulia dam is faced with this rock. The main rock type is a riebeckite quartz syenite. It is a medium-grained, white rock speckled by black amphiboles. It probably belonged to a ring-complex similar to Sabaloka, but not so large, and without the preserved volcanic rocks (Almond 1968).

Jebel Qeili is also a ring-complex, which pierces basement greenschists about 125 km east of Khartoum (Ahmed 1968). It consists of a number of ring intrusives of sodic syenites and granites, and minor remnants of alkali rhyolites and small intrusions of gabbro and dolerite. In general the rock is too coarse to be used for fashioning stone tools, but the jebel has been a burial site and Meroitic inscriptions have been preserved there for the past 2 000 years.

Other ring-complexes are widespread in Sudan (Vail 1972, 1973a) and all are made up of microgranites or syenites. Gabbros are present in some, and relicts of volcanic rocks are preserved in a few. The nearest other ring-complex is at Jebel Mundara in the Butana (Delany 1955); others occur in western Ethiopia, in the valley of the Blue Nile and its tributaries.

4.3.3 *Nubian sandstone formation*

The most important unmetamorphosed sedimentary sequence in Sudan is the Nubian Sandstone Formation. In the central Sudan the formation crops out north of Omdurman and Khartoum and in scattered exposures further south. It is the southernmost extent of a larger area which continues northwards towards Dongola and the Egyptian frontier. Whether the sediments all make up a single stratigraphic unit has yet to be proven, but they are all grouped together on the basis of their lithology and general stratigraphic position.

Outcrops of the Nubian Sandstone Formation within the Gezira are rare, but boreholes and wells have proven the existence of the beds beneath most of the area, and the approximate suboutcrop between Fatisa and Sennar is shown on the map (Figure 1; see also El Boushi & Abdel Salam, chapter 5).

The sediments consist of usually coarse, poorly sorted cross-bedded pebbly sandstones, quartz pebble conglomerates, arkoses, and lesser shales and mudstones. The sandstones are usually cream, white or reddish in colour, and the mudstones range from white to yellow, maroon and reddish; they are fine-grained, and brittle when indurated. The coarse rocks, on the other hand, tend to be soft, friable and easily weathered, although silicification does take place, resulting in a tough enough rock to have been used for hand-axes by early man. The latter are abundant near Omdurman.

The Nubian Sandstone Formation rocks have been used locally as a building stone, and were quarried extensively at Jebel Aulia. They have also been used in Meroitic temples, pyramids and monuments (Shinnie 1967) further north in the vicinity of Shendi. The formation is important as an aquifer and has recently been extensively drilled for water in the Gezira plain and elsewhere.

Fossils are almost entirely absent in the Nubian Sandstone Formation although plant fragments, leaf impressions and silicified fossil wood have been found. Fossil tree trunks and small fragments are common. Some of the latter have been reported from Shabona, where pieces of Nubian Sandstone have been noted.

Due to poor surface exposure, where the more resistant sandstones dominate, the stratigraphic subdivision is not easy and reliance must be placed on subsurface data. Suffice to state here that from lithological and palaeontological evidence it is probable that the Nubian Sandstone Formation, at least in central Sudan, is a fluvial continental deposit, which perhaps accumulated in point bars, delta fans, and flood plains on a relatively flat basement surface of subdued relief. The main direction of transport was from the south and south-east, as evidenced from palaeo-current structures (Kheiralla 1966).

4.3.4 *Tertiary volcanic rocks*

East Africa and Ethiopia have extensive regions in which volcanic outpourings took place from Tertiary up to present times. These events also occurred in many parts of Sudan, although not to the same extent (Francis et al 1973; Medani 1973).

In the area under review basic igneous rocks were emplaced in Nubian Sandstone strata presumably during Tertiary times. Two small outcrops of olivine basalt occur south and west of Omdurman. The former makes up the low jebel of Toriya and is now quarried for road metal (Qureshi, Almond & Sadig 1966; Almond 1968). The other is a small outcrop of hexagonally jointed fine-grained dark green basalt.

Similar rocks are reported from drill hole intersections in Nubian Sandstone Formation between Jebel Aulia and Khartoum, and undoubtedly other buried intrusive masses are present elsewhere in the region.

At Sennar a coarse-grained basalt, referred to as the Sennar teschenite (Whiteman 1971) crops out just below the dam wall, where it is surrounded by Nubian-type sediments.

These basalts and dolerites are minor phases of much more extensive basic and acid volcanism in Ethiopia, but their presence means that one need not go great distances from the Gezira plain to seek such rocks when they have been used by the ancients in their toolmaking. No obsidian, however, occurs *in situ* anywhere nearby in Sudan. The only known locality other than in Ethiopia, is at Jebel Marra in western Darfur Province, and it seems likely that such material when found at the archaeological sites must have come from some hundreds of kilometres away, probably in the highlands of Ethiopia.

There is also no evidence of volcanic activity in Quaternary or Recent times in central Sudan. Jebel Marra was undoubtedly active till quite recently as testified by the hot springs, fumaroles, and unweathered morphology (Hammerton 1966; Vail 1973). Carbon 14 analyses from Deriba crater suggest the last phase of violent activity took place about 4 000 years ago (Francis et al. 1973). Young volcanic features in the Bayuda desert, north-west of Atbara (Almond et al. 1969) also might have been active recently judging from the preservation of the structures. The nearest other active volcanoes must have been in Ethiopia. In that country activity is still present along the Rift zone, and it is highly likely that others, nearer to Sudan, were erupting till quite recent times. They have not, however, been systematically mapped or studied in detail yet.

4.3.5 *Unconsolidated silts and clays of the Gezira plain and Umm Ruwaba Formation type*

By far the most extensive of all the units in the Gezira region of central Sudan are the silts and clays that blanket much of the countryside. Their stratigraphic subdivisions, lithological sequences, precise correlation, and age are problems which remain to be worked out. (A summary of this is given in chapter 5 by El Boushi & Abdel Salam.)

In this outline they will be treated as one mappable unit, pending eventual classification. In the central Gezira region the few hundreds of metres of silts and clays are called the Gezira Formation (Andrew 1948). Similar sediments to the west of the Nile, especially in the type area west of Tendelti, are referred to as the Umm Ruwaba Formation (Andrew & Karkanis 1945), whereas thick unconsolidated fluvial deposits up river from the Dinder, Rahad and Blue Nile river confluences are referred to as the El Atshan Formation (Whiteman 1971). In the east relatively thin, black cotton soils and silts which are characteristic of the Butana plains have been called the Butana Clays. For the present purpose they can all be grouped together since they share the same relative geological position, lithology, and surface expression.

That these beds were transported is evident from their layered nature, and argillaceous character regardless of whether they overlie Nubian Formation sandstones and grits, Basement gneisses and granites, greenschists or igneous ring-complexes.

It seems highly likely that these silts and clays were transported by some early drainage system – a proto Nile perhaps – from the highlands of Ethiopia across the central Sudan plain. There is some evidence of local lacustrine conditions, and perhaps the ridge of Nubian and basement rocks extending across the north of the area acted as a local barrier. It is significant that they do not occur north of an ENE trending line through the latitude of Khartoum, whereas they extend southwards for hundreds of kilometres down the White Nile valley and back up the Blue Nile system to the foothills of Ethiopia. Their age is not precisely known, but they must continue to geologically quite recent times. The distinction of these clays from the recent Nile alluvial beds is not always clear. This problem is further discussed in chapters 6 and 9.

4.3.6 *Wind blown sands, alluvium and superficial gravels*

The youngest cover of the area is made up of sedimentary deposits which are still being formed today or which are of relatively recent origin.

To the west of the White and main Nile, a long, narrow belt of sand occurs, known as Qoz Abu Dulu. This is an ancient streamer of apparent aeolian origin, now fixed, although in disturbed places the sands still shift under the influence of prevailing winds. Seif dunes are common, and in places crescentic or transverse forms are present. Between Tendelti and Kosti this dune belt widens out, and there is a possibility of an interplay between the Nile drainage and the sand deposit. The possibility has been considered that this sand belt might even represent a levee course of an ancient Nile stream, but lacking detailed investigation this idea cannot be supported.

The sands are usually of frosted, rounded grains of quartz. Near the surface the sands are stained red by surface oxidation, but in depth are usually white. They contain too much iron to be useful as pure quartz sands, but being absorbant they are good, but shallow and temporary aquifers. Between the dunes mud flats are encountered, which in part grade into the river alluvium in appropriate places.

East of the Nile, particularly around the northern side of the Butana, a number of remarkable gravel deposits occur. Much of the basement rocks, and even the Nubian Sandstone Formation of Sudan, are covered by loose, lag gravels, usually almost entirely composed of quartz fragments. In the Butana these gravels mark the contact between the Butana clays and the Nubian sandstones. Some are influenced by the wind, but most appear to be residual in nature.

The valleys of the three Nile branches and several of the smaller tributaries have well marked alluvial deposits. These are usually silts and clays, but include sand bars and mud flats, many of which are in active modification with changing seasons. Ancient river terraces have been recognized along the

Nile, and their precise elevation, location and nature are of immense importance to agriculture (chapter 7). The alluvial deposits as shown on the accompanying map (Figure 1) are very generalized. Detailed soil mapping, profile levelling and pedology will provide a more accurate picture in the Gezira area.

The alluvium along the Nile is of importance to the inhabitants for, away from the Gezira, it is the main area available for cultivation and irrigation and is the site of some of the earliest settlement.

4.4 GEOLOGICAL HISTORY OF THE AREA

The history of the pre-Nubian basement rocks in the central part of Sudan can only be hinted at, since regional studies have only just started on this problem.

It seems probable that ancient sedimentary rocks underwent tectonothermal modification in at least one major period of orogeny. Probably as part of this activity large granitic masses were emplaced, adding further to the metamorphism of the country rocks. It might be surmised that following this was a period of erosion followed by a new depositional phase, dominated by volcanic activity, now represented by the greenschist group. Further folding, and probable granite emplacement, took place which resulted in the dominant structures now seen in the basement rocks.

From exposures in the Gezira area it is evident that the gneisses trend in a broad band north-eastwards from the vicinity of Kosti to Jebel Qeili. In the extreme south the rocks trend north-west.

These basement rocks are of quite unknown age, but it can be surmised that the whole area was subjected to the late effects of the Pan African orogenic event (Kennedy 1964), since such isotopic age measurements as are available indicate an event of about 600 million years in age (Vail & Rex 1970).

Presumably following this was a period of quiescence and erosion, during which the sub-volcanic igneous complexes were emplaced at Sabaloka and elsewhere. The age of Sabaloka is not yet unambiguous but isotopic results suggest it must be older than about 540 million years (Almond 1967) and Briden (1973) suggests it may have been closer to 710 million years in age.

There then followed a considerable hiatus, during which time any sedimentary rocks that might have been deposited were removed by erosion, until the sands and clays of the Nubian Sandstone Formation began to accumulate. This must have been during Mesozoic times, and it is considered that they were deposited under aqueous conditions across a broad plain from high ground in the south.

Sedimentation has continued sporadically since that time when silts and clays became dominant and were spread out across plains of older rocks. Tothill (1946) considered the clays of the Gezira plain to have been fluvial deposits carried in from the south, a conclusion supported by Williams & Adamson (1976, 1980).

Intermittent volcanic activity occurred, but died out in central Sudan in early times, and the area became dominated by climatic fluctuations which brought in wind blown sands from the north and alluvial deposits from the south. These conditions have continued since the early Pleistocene until the present day, and the history of the more recent geological events shows that central Sudan has long been a zone of transition between arid desert and rainland savannah, yet with a plentiful supply of water in the rivers.

4.5 REFERENCES

Abdel Salam, Y. 1966, The ground water geology of the Gezira. M.Sc. thesis, Univ. Khartoum.

Ahmed, F. 1968, The geology of the Jebel Qeili, Butana and Jebel Sileitat-es-Sufur igneous complexes, Nile valley, central Sudan. M.Sc. thesis, Univ. Khartoum.

Almond, D.C. 1967, Discovery of a tin-tungsten mineralization in northern Khartoum Province, Sudan. *Geol. Mag.* 104:1-12.

Almond, D.C. 1968, Petrology of the basalt at Jebel et Toriya, Khartoum. *Sudan Notes and Records* 67:141-150.

Almond, D.C. 1971, Ignimbrite vents in the Sabaloka cauldron, Sudan. *Geol. Mag.* 108: 159-176.

Almond, D.C., F. Ahmed & B.E. Khalil 1969, An excursion to the Bayuda volcanic field of northern Sudan. *Bull. Volcanologique* 33(2):544-565.

Andrew, G. 1948, The geology of the Sudan. In J.D. Tothill (ed.), *Agriculture in the Sudan.* London, Oxford University Press.

Andrew, G. & G.Y Karkanis 1945, Stratigraphical notes, Anglo-Egyptian Sudan. *Sudan Notes and Records* 26:157-166.

Berry, L. & A.J. Whiteman 1968, The Nile in the Sudan. *Geogr. J.* 134:1-37.

Briden, J.C. 1973, Palaeomagnetic estimate of the age of the Sabaloka Complex, Sudan. *17th Ann. Rep. Inst. Afr. Geol.* Univ. Leeds.

Dawoud, A.S. 1970, A geological and structural study of the area south-east of Qerri Station, eastern Sabaloka, northern Khartoum Province. M.Sc. thesis, Univ. Khartoum.

Delany, F. 1952, Geological map of Sudan. Sheet 55 Khartoum, scale 1:1 million. *Geol. Surv. Sudan.*

Delany, F. 1953, Geological map of Eastern Khartoum Province, scale 1:250 000. *Geol. Surv. Sudan.*

Delany, F. 1955, Ring structures in the northern Sudan. *Eclogae Geol. Helvetiae* 48:133-148.

Delany, F. 1958, Observations on the Sabaloka Series of the Sudan. *Trans. Geol. Soc. S. Afr.* 61:111-124.

El Boushi, I.M. 1972, The shallow ground water of the Gezira formation at Khartoum and the northern Gezira. *Sudan Notes and Records* 53:154-163.

Francis, P.W., R.S. Thorpe & F. Ahmed 1973, Setting and significance of Tertiary-recent volcanism in the western Sudan. *Nature Phys. Sci.* 243 (124):30-32.

Hammerton, D. 1966, Hot springs and fumeroles at Jebel Marra, Sudan. *Nature* 209 (5030):1290.

Howie, R.A. 1958, African charnockites and related rocks. *Bull. Serv. Géol. Congo Belge* 8(2):1-14.

Hunting Technical Services Ltd. 1964, Roseires soil survey. *Rept. to Min. of Agric., Republ. of Sudan.*

Kabesh, M.L. 1967, Petrographic notes on some charnockitic rocks from the Sudan. *Bull. Soc. Géogr. d'Egypte* XL:157-172.

Kennedy, W.Q. 1964, The structural differentiation of Africa in the Pan-African (± 500 m.y.) tectonic episode. *8th Ann. Rep., Inst. Afr. Geol.* Univ. Leeds:48-49.
Kheiralla, K.M. 1966, A study of the Nubian sandstone formation of the Nile valley between latitudes 14°N and 17°42′N, with reference to the groundwater geology. M.Sc. thesis, Univ. Khartoum.
Medani, A.H. 1973, Tertiary basalts of eastern Sudan. *Nature Phys. Sci.* 242(116):61-62.
Omer El Badri 1972, Sediment transport and deposition in the Blue Nile at Khartoum, flood seasons 1967, 1968 and 1969. M.Sc. thesis, Univ. Khartoum.
Qureshi, I.R., D.C. Almond & A.A. Sadig 1966, An unusually shaped basalt intrusion in the Sudan. *Bull. Geof. Teor. Appl.* 3:151-160.
Sadig, A.A. 1969, A gravity and magnetic survey of the Sabaloka area (Sudan) with details of the ring complex. M.Sc. thesis, Univ. Khartoum.
Shinnie, P.O. 1967, *Meroe, A civilization of the Sudan.* London, Thames and Hudson.
Tothill, J.D. 1946, The origin of the Sudan Gezira clay plain. *Sudan Notes and Records* 27:153-183.
Vail, J.R. 1972, A ring-complex igneous province in northern Sudan. In T.E. Dessauvagie & A.J. Whiteman (eds), *Afr. Geol. Ibadan 1970:*163-177.
Vail, J.R. 1973a, Distribution of igneous ring-complexes in Sudan and vicinity. *17th Ann. Rept., Inst. Afr. Geol.:*27-30. Univ. Leeds.
Vail, J.R. 1973b, Jebel Marra, a dormant volcano in Darfur province, western Sudan. *Bull. Volcanologique* 36(1):251-265.
Vail, J.R. & D.C. Rex 1970, Tectonic and geochronological studies in Sudan. *14th Ann. Rept., Inst. Afr. Geol.:*44-47. Univ. Leeds.
Whiteman, A.J. 1971, *The geology of the Sudan Republic.* Oxford, Clarendon Press.
Williams, M.A.J. & D.A.Adamson 1976, The origins of the soils between the Blue and White Nile rivers, central Sudan, with some agricultural and climatological implications. Econ. and Soc. Res. Council, Natl Council for Res. Sudan, Occasional Paper 6, Khartoum.
Williams, M.A.J. & D.A.Adamson, 1980, Late Quaternary depositional history of the Blue and White Nile rivers in central Sudan. In: M.A.J.Williams & H.Faure (eds), *The Sahara and the Nile.* Balkema, Rotterdam, 281-304.

ISMAIL M.ELBOUSHI
ERTEC, Long Beach, Ca., USA
YASSIN ABDEL SALAM
University of Khartoum, Sudan

5

Stratigraphy and ground-water geology of the Gezira Plain, central Sudan

5.1 INTRODUCTION

'Gezira' (Arabic: 'island') often denotes an area fertile for agriculture, which is appropriate for what has long been one of the largest irrigation areas in the world. The Gezira plain is mantled by a series of sands, silts, and clays. Differences in physical and chemical properties of the Gezira soils and sediments influence groundwater quality and soil fertility, and are a legacy from the past. The presence of both fertile soils and potable supplies of groundwater has a profound influence on the present distribution of population, an influence which may go back several millennia. This chapter deals with the distribution, occurrence, and stratigraphy of the sediments on top of the Basement Complex rocks, as well as with their groundwater. At a time when the Sudan is being mooted as the future granary of the Arab world, a knowledge of its water resources is as relevant to modern man as it was, albeit on a smaller scale, to the early inhabitants of this area.

5.2 STRATIGRAPHY OF THE SEDIMENTS IN THE GEZIRA AREA

The sediments occurring in the Gezira comprise the Gezira Formation which rests unconformably on the Nubian Sandstone Formation (Abdel Salam 1966; Whiteman 1971), which in turn lies unconformably on the Basement Complex rocks (Table 1).

5.2.1 *Nubian Sandstone Formation*

The Nubian Sandstone Formation occupies about 28 % of the surface area of the Sudan and is the main sedimentary sequence and aquifer in the Sudan. It crops out at few localities in the Gezira. These localities are mainly at Jebel Aulia, Qoz Barik, and Qoz Beida. It is also found on the banks of the White Nile at El Geteina, and in the Blue Nile valley at Alti, El Masid and Sennar.

Table 5.1. Summary of stratigraphy of the Gezira.

Sediments	Locality	Lithology	Thickness (m)	Age	Depositional environment	Hydrogeological characteristics: Depth to water (m)	Aquifer thickness (m)	T (m^2/day)	S	Salinity (ppm)
Nile and wadi deposits				Cainozoic	Fluvial	–	–	–	–	–
—— UC ——										
Dune sands	A	Quartz sand	1-15		Aeolian	–	–	–	–	–
—— UC ——										
White Nile terrace deposits	B	Silt, clay,	4-9		Fluvio-lacustrine	–	–	–	–	–
—— UC ——										
Dune sands	C	Quartz sand	? up to 10		Aeolian	–	–	–	–	–
—— UC ——										
Clay Member	D	Clay, fine sand and silt	1.5-180		Fluvio-lacustrine	4.0-50	18-25	380 to 690	25×10^{-4} to 46×10^{-7}	300 to 4 200* (mean 420)
Sandy Member		Sand and gravel								
Mungata Member		Clay, sand and gravel								
—— UC ——										
Nubian Sandstone Formation	E F G	Conglomerate, sandstone, mudstone	75 240 120	Mesozoic	Fluvio-lacustrine	up to 76	up to 42	185 to 100	2×10^{-2} to 27×10^{-9}	400 to 600

Localities:
A: El Geteina area (14°15′N to 15°15′N); B: Khartoum to Melut; C: Kosti area; D: Ghubshan (14°14′N, 33°02′E); E: Jebel Aulia (15°16′N, 33°30′E); F: Qoz Beida and Qoz Barik; G: 13°55′N, 33°13′E; T: transmissivity; S: storage coefficient; *Locally up to 32 200.

Elsewhere in the Gezira it is covered by the Gezira Formation and superficial deposits of variable thickness.

In the northern part of the Gezira the Formation consists of conglomerates, grits, sandstone, kaolinized feldspathic sandstones, sandy mudstones, and mudstones. The dominant rock type is medium- to fine-grained sandstone with intercalations of mudstone and kaolinized feldspathic sandstone. When the cement is argillaceous the rocks are often soft, but silica and iron oxide cements give rise to hard sandstones.

In the south central Gezira the Nubian Sandstone Formation generally contains less mudstone than in the northern Gezira, especially at Manaqil where the Nubian Formation has been subjected to local erosion and deposition thus producing thick layers of intraformational conglomerates. At Sennar, the Formation contains more sandy and pebbly horizons which attain a thickness of 38 m. The absence of outcrops south of Sennar and the decrease in grain-size in a northwest direction, together with the palaeocurrent evidence given by Kheiralla (1966), suggest that the formation in the Gezira was derived from sources south of Sennar. The presence of intraformational conglomerates and the high percentage of clays and silts in the sandstones suggest that the deposition took place under aqueous to subaqueous conditions from fast-flowing rivers (Whiteman 1971).

The Nubian Formation is devoid of animal fossils but contains fossil plants represented by silicified wood and leaf remains, many of which are identifiable at generic level (Whiteman 1971). The Nubian Formation is probably Mesozoic in age (Whiteman 1971).

5.2.2 *Gezira Formation*

The Gezira Formation, which rests unconformably on the Nubian erosional surface, was first recognized by Andrew (1948) when he used this name for the unconsolidated clays, silts, sands, and gravels in the Gezira area. The precise distinction between weathered Nubian Formation sediments and those of the overlying Gezira Formation is not always easy to detect, particularly from borehole data. There is clear need for a re-appraisal of the Gezira Formation and its constituent units, and the choice of El Ghubshan as the type locality may prove in time to be debatable. Until a great deal more field evidence is forthcoming to resolve these issues, we will accept Abdel Salam's (1966) provisional interpretations, bearing in mind the above reservations. Abdel Salam (1966) gave El Ghubshan (14°14′N 33°02′E) as the type locality. The Formation there attains a thickness of over 60 m. It consists of 9 m of coarse gravelly sand at the base that passes upwards into clayey sands 35 m thick. The upper part of the Formation consists of 17 m of dark clays and silts locally known as cotton soil. Though the Gezira Formation is characterized by rapid facies changes, Abdel Salam (1966) recognized three subdivisions. These are: Upper Clay Member, Lower Sandy Member and Mungata Member.

5.2.2.1 *Upper Clay Member.* This unit passes downward by gradation into the Lower Sandy Member. It consists mainly of clays with occasional patches

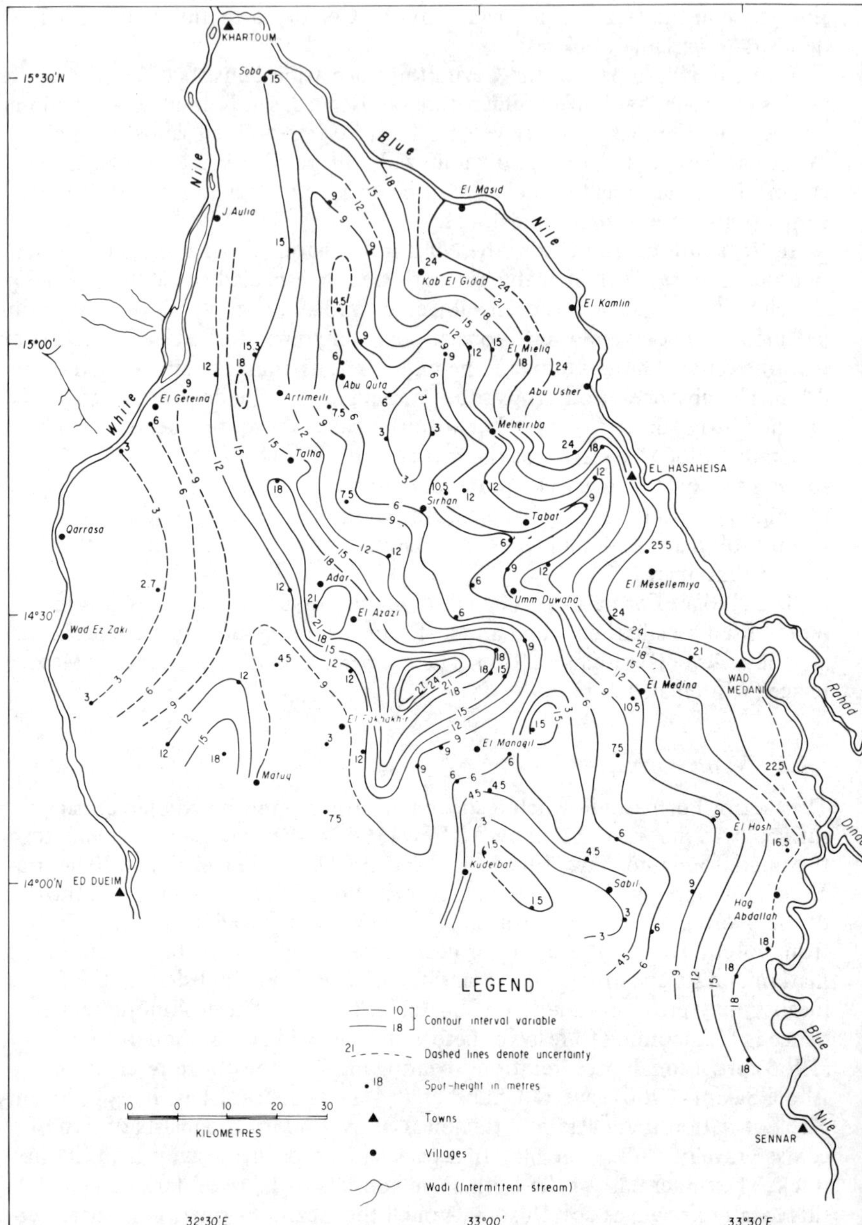

Figure 5.1. Isochore map of the Upper Clay Member, Gezira Formation, (Data from records of the Ministry of Natural Resources and Rural Water Development, Ground Water Section).

of sand and pockets of quartz pebbles. Kankar nodules occur frequently in localized areas in this Member. The Upper Clay Member varies in thickness from 27 m east of El Fakhakhir to less than 3 m close to the White Nile (Figure 1). Near the Blue Nile it is 24 m thick, but decreases westwards to 3 m at the centre of the Gezira (Figure 1).

5.2.2.2 *Lower Sandy Member.* The Lower Sandy Member consists mainly of sands with occasional gravel, silt, and clay lenses. The sand is often clayey and calcareous. The gravel lenses vary in thickness from a few centimetres to over 30 m. Individual lenses of gravel sometimes extend for a distance of 16 to 24 km and usually thin out towards the Blue Nile. The thickness of this Member varies from 20 m at Amara (14°44′N 33°22′E) to more than 58 m at Umm Busha (14°45′N 32°59′E).

5.2.2.3 *Mungata Member.* The Mungata Member was recognized by Abdel Salam (1966). It is a subsurface unit which can be identified only from well data in the south central part of the Gezira around Mungata from which the name was derived. It occurs at Wad Adam (14°21′N 32°37′E) and west of El Manaqil. In general the Mungata Member is predominantly clayey and consists of coarse arkosic sand, clayey grits, and sands. The percentage of clay varies from 70 to 80 %. The clay is often green, but may be grey, brown, or buff. Fish-bone fossils have been recorded by the Sudan Geological Survey Officers from boreholes at Tarfa (14°18′N 32°37′E) at a depth of 66 m and at El Geleiz (14°32′N 32°39′E) at a depth of 44 m. At the type locality, the Mungata Member has a thickness of more than 111 m.

5.2.3 *Lithology of the Gezira Formation*

In general the Gezira Formation consists of clay, sand, and gravel. Detailed description of these sediments is given below.

5.2.3.1 *Clay.* According to Ruxton (1956) the clays of the Gezira Formation are alkaline, dark in colour, and low in organic matter. They contain a high proportion of undecayed minerals such as calcic plagioclase, titan-augite, hornblende, and brown biotite flakes (Andrew 1948). The clays vary from dark grey at the top to brownish-grey at depth. Figure 2 shows the mean proportion of sand to clay in the Gezira Formation. From this figure it can be seen that the clay is mainly present in the Fakhakhir-Mungata area. It occurs also in small isolated pockets of erratic distribution extending from the area west of El Mieliq to El Hosh (Figure 2). The clay content in these small pockets is about 60 %, but may reach 80 % in the Mungata-Fakhakhir area.

5.2.3.2 *Sand.* The sand is composed mainly of angular and subangular quartz grains. The grain-size is predominantly fine to medium, and the sorting is poor. Coarse fragments are often present and in places the sand contains considerable quantities of clay. When washed the sand grains are whitish-yellow or pink. Occasionally the sand is micaceous, especially in the upper parts of

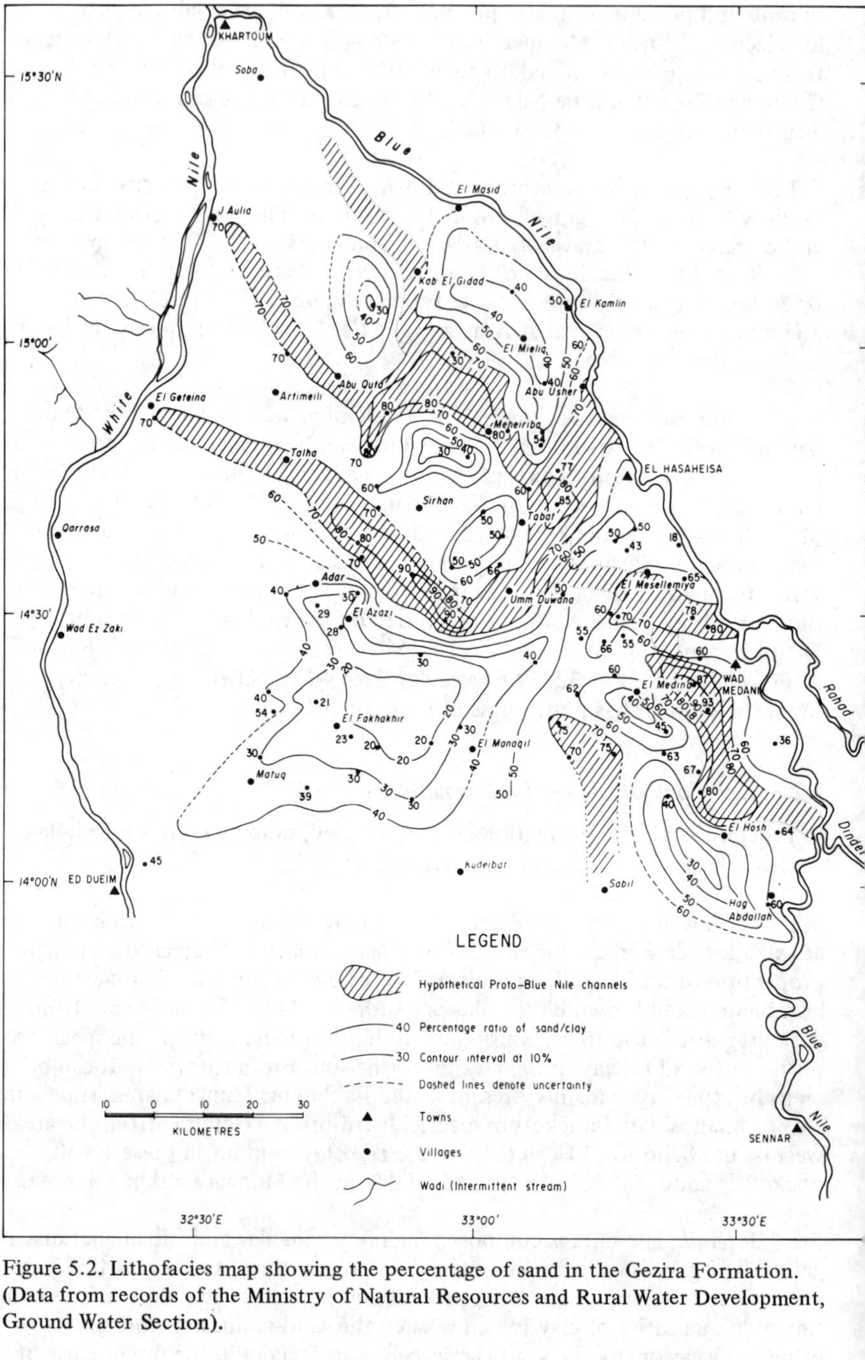

Figure 5.2. Lithofacies map showing the percentage of sand in the Gezira Formation. (Data from records of the Ministry of Natural Resources and Rural Water Development, Ground Water Section).

the Gezira Formation where concentrations of ferromagnesian minerals sometimes occur. The sand is friable, but in places it is cemented by calcium carbonate. In general the Gezira Formation is predominantly sandy, as evidenced from the map showing the ratio of sand to clay (Figure 2). The content of sand is more than 60 % close to the Blue Nile at El Medina, and is over 85 % at and around Tabat, Abu Quta, and Artimeili (Figure 2).

5.2.3.3 *Gravels.* The gravels in the Gezira Formation consist of quartz pebbles with scattered chert fragments, iron pisolites, kankar nodules, agates, basalts, and fragments of metamorphic and igneous rocks. The quartz pebbles are usually white or yellow and are similar to the quartz commonly found in the veins of the Basement Complex in the southern parts of the Gezira. They are subangular to rounded, and average 2 to 3 cm in diameter. The iron pisolites are reddish-brown with dark grey centres. They are formed of limonitic material cementing fine sandy particles. The pisolites are rounded and separate, but sometimes two or three may be connected to each other. Their mean diameter is 5 cm.

The kankar nodules are yellowish-white at the surface, with brownish- to dark grey iron or quartz centres. They are formed predominantly of calcium carbonate with a few fine sandy fragments. The kankar nodules are dendritic and rounded. Sometimes many of them may be connected to each other. Their size ranges from 3 to 5 cm. Sheets of calcium carbonate crop out in cliffs along the Blue Nile southeast of Khartoum. At El Gereif, Soba, and Alti such sheets appear to be forming today.

5.2.4 *Thickness of the Gezira Formation*

The thickness of the Gezira Formation varies from 1.5 m south of Manaqil to more than 180 m near the centre of the Gezira (Figure 3). Where the Formation is thin it consists mainly of clays as at Soba, where it comprises 15 m of clay resting on the Nubian Sandstone Formation. In the southern Gezira east of Wad Medani the Formation is predominantly formed of clays resting on the Nubian Formation. It varies there from 1.5 m at Kudeibat and in the area east of El Manaqil to 60 m at Umm Duwana and Fakhakhir (Figure 3).

Figure 3 is an isochore map of the Gezira Formation and reveals the topography of the sub-Gezira erosional surface. From this map it is clear that there is a ridge extending from El Hosh to Fakhakhir. This ridge has been recognized before by Abdel Salam (1966), who called it the Manaqil ridge. The Gezira Formation is thin above this ridge, which is mainly composed of Nubian Sandstone. The northern extension of the Manaqil ridge is found at Umm Duwana (Figure 3).

Another feature of the sub-Gezira erosional surface is the presence of a basin with thick Gezira sediments at Kab El Gidad. This basin extends in a southwesterly direction passing by Abu Quta, Talha, and may end near Matuq (Figure 3).

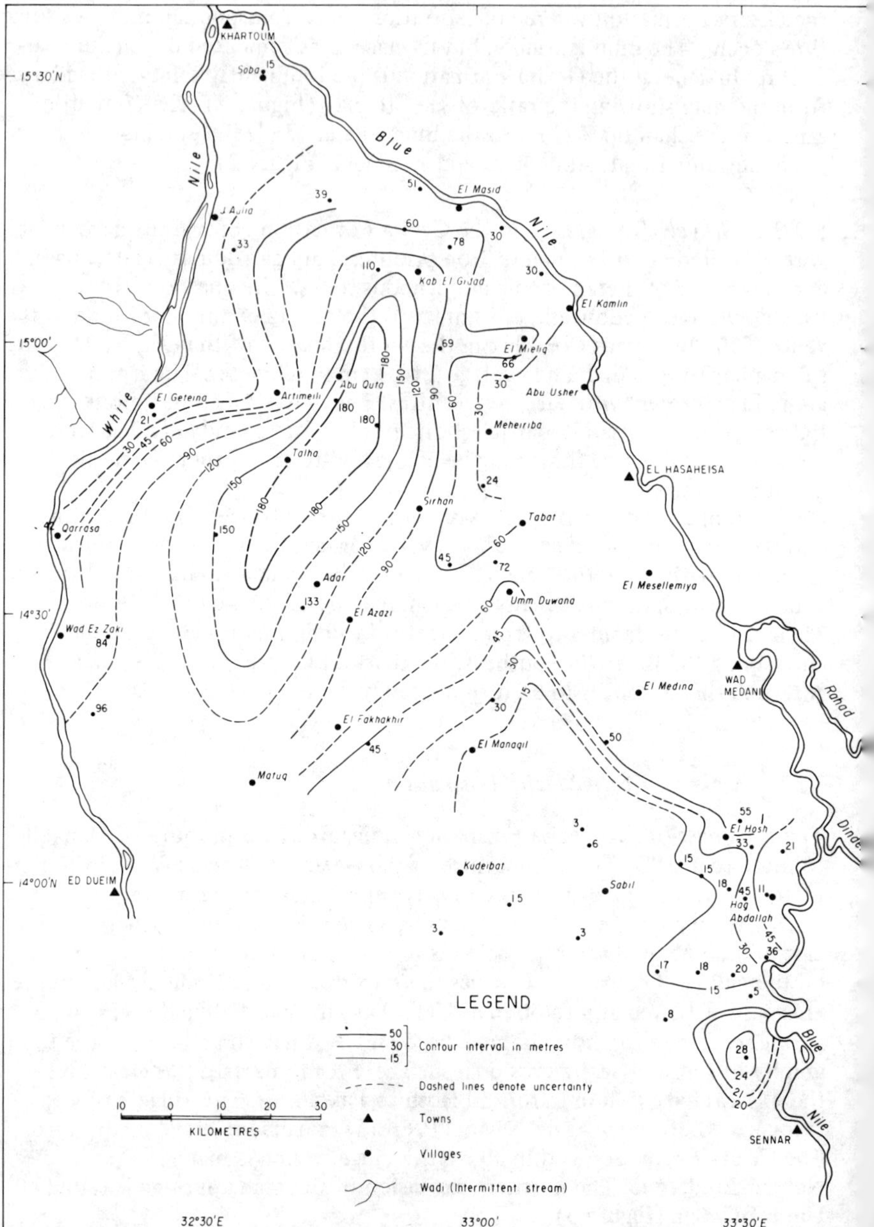

Figure 5.3. Isochore map of the Gezira Formation. (Data from records of the Ministry of Natural Resources and Rural Water Development, Ground Water Section).

5.2.5 *Age of the Gezira Formation*

The age of the Gezira Formation is not definitely known, although the upper part of the Formation is probably of Pleistocene and Recent age (Arkell 1949; Whiteman 1971). Tothill (1946, 1948) and Jewitt (1955), using evidence from fossil mollusca, heavy minerals, soil chemical and physical composition, and prehistoric stone artifacts, have argued that throughout much of the Gezira the upper clays were deposited between about 50 000 and 10 000 years ago by floods from the Blue Nile. The Gezira clays overlie clays and kankar gravels containing advanced Levalloisian type implements (Lacaille 1951). Records of implements near Khartoum led Andrew (1948) to suggest that the clay of the plain was probably post-lower Palaeolithic and pre-Neolithic in age. Sand and gravel penetrated during the excavations of the foundation for the Blue Nile Bridge yielded an elephant tooth, considered by Andrews (1912) as Mid-Pleistocene. The age of the Gezira clays is discussed further in chapters 7 and 9.

5.2.6 *Origin of the Gezira Formation*

Much has been written but little is yet known except in very general terms concerning the provenance of the Gezira Formation. Shukri (1950), Worrall (1956) and Badri (1972) described the mineralogy of the Gezira clays and recorded calcic plagioclase, augite, hornblende, biotite and olivine. This assemblage is similar to that being transported by the modern Blue Nile and is a volcanic assemblage derived from the Ethiopian uplands. Typical Blue Nile minerals such as pyroxene do not seem to occur in the sands and gravels but are common and characteristic of the Gezira clays (Whiteman 1971).

The sands and gravels in the Gezira Formation were probably derived from the Nubian Formation near Sennar and Gedaref, the Adigrat Formation in Ethiopia, and the Basement Complex rocks which crop out southeast and west of Sennar. Accordingly the Gezira Formation is polygenetic with much of the clay fraction having been swept in by rivers draining the Ethiopian volcanic plateau, and with the coarse fraction probably derived from the weathering of the Nubian Formation and the Basement Complex in Sudan.

5.2.7 *Deposition of the Gezira Formation*

Two hypotheses have been proposed to account for the origin of the Upper Clay Member of the formation: Lawson (1927) and Ball (1939) postulated a lake to account for the origin of the Gezira clays, but the available archaeological, geomorphological, and hydrological evidence does not support a former 'Lake Sudd' on the scale proposed by them.

Tothill (1946), Berry (1964), Williams (1966) and Whiteman (1971) have attributed the formation of the Gezira clay member to overbank flooding by the Blue Nile and its tributaries. Large-scale meanders of the Blue Nile have been recorded from aerial photographs in the Gezira (Williams 1966 and chapters 7 and 9). The fossil gastropods collected by Tothill (1946) and by

Williams (1966) confirm the reality of overbank flooding. The clay member is thick all along the Blue Nile but it becomes thin towards the centre of the Gezira (Figure 1). It is also thick in a basin located between Talha, Adar, and El Azazi (Figure 1). Accordingly it is possible to suggest that the deposition of the Upper Clay Member took place by overbank flooding from the present channel of the Blue Nile and from a proto-Blue Nile running in the basin of Talha and Adar. The climate under which the clays were deposited was probably somewhat wetter than at present (see chapters 7 and 9). Gypsum crystals were precipitated in the clay during the long period of desiccation which then ensued (Andrews 1948).

The Lower Sandy Member was deposited in fans and deltas by now buried distributary channels of the former Blue Nile. Conditions similar to those prevailing in the present-day Gash delta (Swan 1956) may well have existed over the Gezira. Buried channels and possible distributaries of the Blue Nile may be discernible from the subsurface data (Figure 2). On the assumption that the coarse fraction forms the channel bed-load (Williams 1968a), 70 % sand was chosen as a criterion to identify the presence of buried channels in the Gezira (Figure 2).

The deposition of the Gezira Formation was initiated by warping along a northeast-southwest axis associated with Tertiary volcanicity. Sediments accumulated in a subsiding basin so that at Wad Adam in the southern Gezira more than 110 m of the Gezira Formation accumulated and bedrock was depressed more than 80 m relative to the sill height at Sabaloka north of Khartoum (Abdel Salam 1966).

5.3 GROUNDWATER IN THE GEZIRA AREA

The groundwater has played and still plays a major role in both ancient and recent settlements in the Gezira. Shallow wells were dug by hand long ago and formed the main source of water. Recently, many deep wells were drilled to supplement the ancient and modern shallow wells. However, for economic reasons, shallow wells are abundant in the Gezira.

There are two main water-bearing formations in the Gezira. These are the Nubian Sandstone and Gezira Formations. Both aquifers contain large quantities of water. Where the Gezira Formation is thin (20 m or less) and clay-rich, the main aquifer is in the underlying Nubian Sandstone Formation. Very often the two aquifers are interconnected and their waters are mixed. Most of the shallow wells tap the Gezira aquifer. The deep wells which penetrate both aquifers tap the two water-bearing formations unless the upper Gezira aquifer is sealed because it contains water of poor quality.

5.3.1 *The Nubian Sandstone aquifer*

The Nubian Sandstone Formation is the main aquifer in the Sudan, but in the Gezira area it occurs at greater depths and consequently is not as well exploited as in other parts of the Sudan. Groundwater is found under semi-artesian conditions owing to the presence of the Gezira Formation and its clays on top of the Nubian aquifer.

The water-bearing beds consist mainly of sandstones and conglomeratic sandstones. The sandstones consist in general of fine- to medium-grained angular quartz particles with a moderate degree of sorting. The average saturated thickness of the Nubian Formation is 42 m (Abdel Salam 1966). The depth to water where wells are definitely penetrating this aquifer is variable. In the southern part of the Gezira and west of Sabil where the Gezira Formation is only 1.5 m thick, the depth to water in the Nubian aquifer is 76 m (Figure 4).

Pumping tests conducted by Geotechnica Company in wells tapping the aquifer revealed transmissivity values ranging from 185 to 585 m^2/day (15 000 to 47 000 gal/day/ft) and storage coefficients varying from 2 x 10^{-2} to 1.7 x 10^{-9}. Such values indicate that the Nubian aquifer is under non-confined to confined conditions. The transmissivity values also show that the Nubian aquifer is a good aquifer containing large quantities of water.

5.3.2 *The Gezira aquifer*

The Gezira aquifer is the main aquifer in this area. The groundwater occurs under semi-artesian to artesian and leaky conditions, as a result of impervious layers of clay in the Gezira Formation (El Boushi 1972). The main water-bearing stratum is the Lower Sandy Member which consists of sinuous interconnected sand bodies. The sands are often clayey and pebbly, but in general they consist of fine, medium and coarse sands. According to Abdel Salam (1966) the thickness of the aquifer ranges from 18 to 25 m. A study in the northern Gezira revealed that the groundwater level is high near the Blue and White Nile rivers and declines in elevation towards a central trough south of Abu Quta (El Boushi 1972).

Generally the depth to water varies from 4 to 12 m near the rivers and reaches up to 50 m in the central trough (Figure 4). Figure 4 shows that the depth to water decreases from Sennar northwards to El Masid, and also indicates that there are two main troughs. The smaller one is situated at El Meiliq where the depth to water reaches about 30 m. The second and largest trough coincides with the one discussed by Abdel Salam (1966). It is present between Talha, El Azazi, and Fakhakhir. The maximum depth to water in this trough is 50 m at El Fakhakhir. Figure 4 confirms an earlier observation by El Boushi (1972) relating to the close parallelism between the alignment of the Blue and White Nile rivers and the groundwater contours. Such a relationship indicates that both Niles are directly connected to and replenish the Gezira aquifer. The troughs formed in the central part of the Gezira represent windows of replenishment to the Nubian aquifer. If water did not leak to augment the Nubian water these troughs would have been filled long ago.

Pumping tests conducted by Geotechnica Company in the Gezira gave transmissivity values for the Gezira aquifer ranging from 385 m^2/day to 535 m^2/day (31 000 to 43 600 gal/day/ft). The storage coefficient of the Gezira aquifer determined by Geotechnica varied from 2.5 x 10^{-4} to 4.8 x 10^{-7},

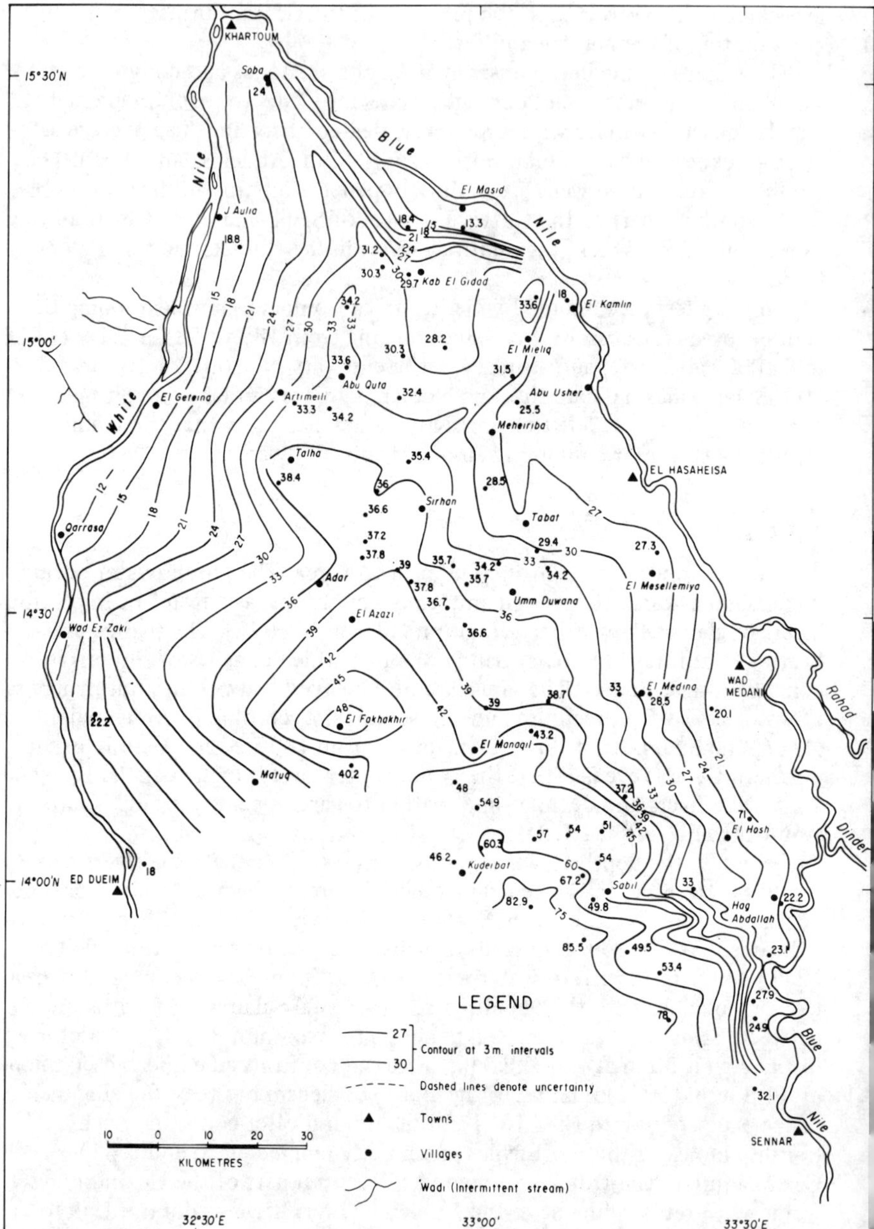

Figure 5.4. Contour map of equal depth to ground water in the Gezira, central Sudan. (Data from records of the Ministry of Natural Resources and Rural Water Development, Ground Water Section).

which indicates that the Gezira aquifer is a good aquifer (de Wiest 1965:171) where water occurs under semi to complete confinement.

After analysing pumping test data, El Boushi (1972) suggested that the Gezira aquifer is a leaky artesian one. The presence of the major trough in the centre of the Gezira aquifer confirms that the leakage takes place and that the water percolates downward to the Nubian aquifer.

A comparison between the transmissivity and storage coefficient values of the Gezira and Nubian aquifers (Table 1) reveals that the Gezira aquifer is more efficient than the Nubian aquifer. This conclusion differs from earlier ideas about these aquifers. The Gezira aquifer is superior to the Nubian aquifer for two main reasons: a) the Gezira aquifer contains well-sorted medium sands in contrast to the ill-sorted argillaceous gravels of the Nubian aquifer; and b) little or no diagenesis has taken place in the Gezira Formation, unlike the well-cemented and indurated Nubian Formation.

5.3.3 *Chemical quality*

The groundwater of the Gezira aquifer has a variable mineral content. The chemical quality varies from good water with a low mineral content to extremely poor and saline water which is unfit for human and animal consumption or agricultural use. The salinity is low close to the Blue Nile where it varies from 300 to 500 ppm, but increases away from the Blue Nile reaching 32 200 ppm 16 km north of Abu Quta and 4 200 ppm near El Azazi and Adar (Figure 5). The salinity is 1 000 ppm near the White Nile but increases eastwards towards the centre of the Gezira.

The Nubian and the Gezira aquifers are definitely connected. Consequently their waters are mixed and it is difficult to separate them into two different types of water at least in the Gezira area. However, northeast of Sennar in the southern part of the Gezira groundwater occurs uniquely in the Nubian Formation and is more or less representative. This water is of good quality with a low mineral content and total dissolved solids varying from 400 to 600 ppm.

The low salinity of the Gezira aquifer close to the Blue and White Nile rivers supports the idea that recharge to the aquifer takes place from these two rivers. As water moves away from the recharge zones it dissolves the more soluble salts from the sediments. Consequently salinity increases to a maximum at the trough in the centre of the Gezira. Prolonged solution by this slowly moving or stagnant groundwater of soluble salts in the surrounding sediments which contain gypsum and calcium carbonate eventually leads to a high salt concentration in the central and other troughs in the Gezira (El Boushi 1972). The presence of gypsum and carbonate in these sediments may itself imply that saline waters accumulated in an internal drainage basin in the past.

Four different chemical types of groundwater can be recognized in the Gezira Formation (Abdel Salam 1976). They are:

1. the calcium bicarbonate group,
2. the sodium bicarbonate group,

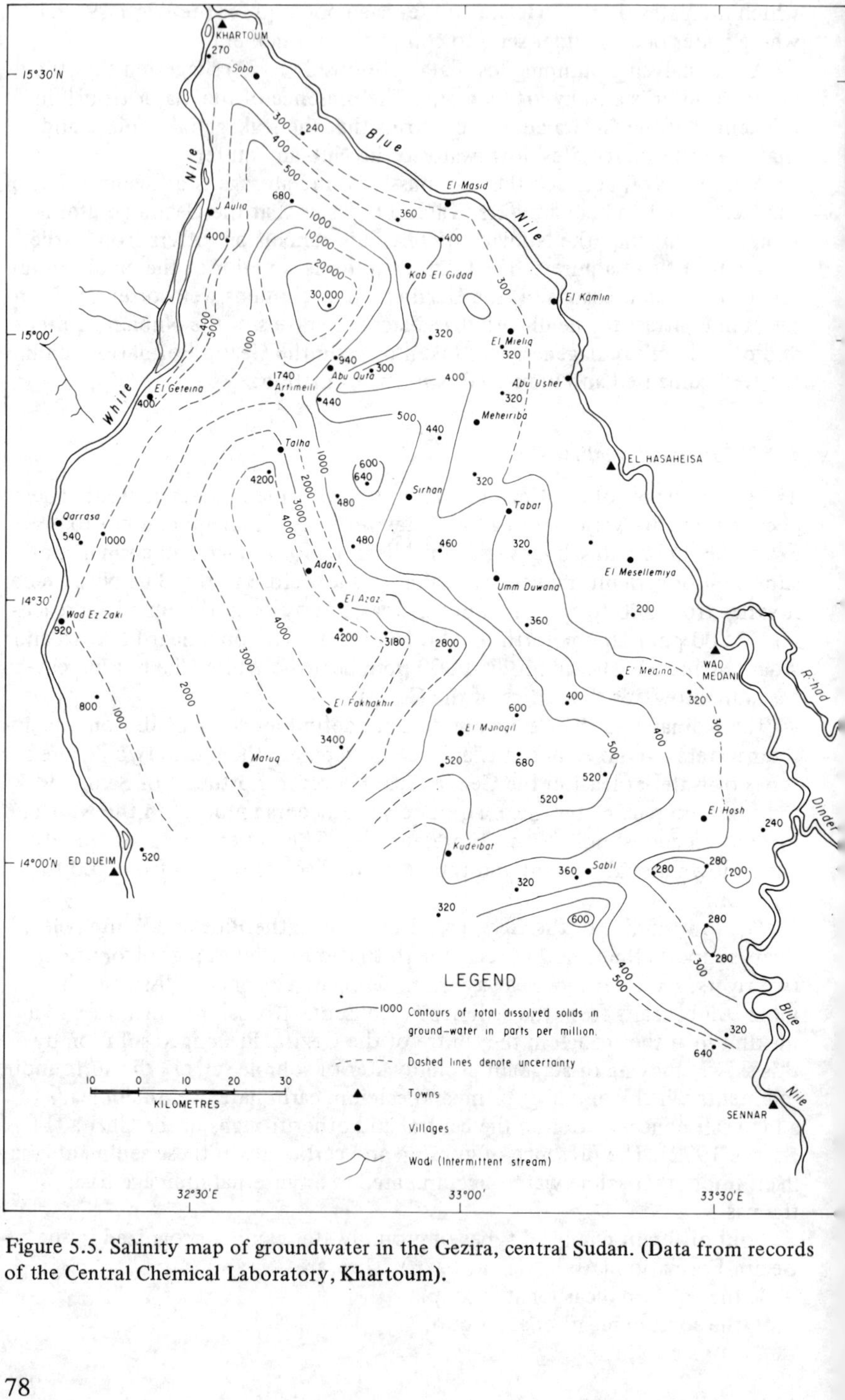

Figure 5.5. Salinity map of groundwater in the Gezira, central Sudan. (Data from records of the Central Chemical Laboratory, Khartoum).

3. the sodium chloride group,
4. the sodium sulphate group.

The calcium bicarbonate group occurs in a zone which fringes the Blue Nile river. The average width of this zone is approximately 7 km. The zone passes gradually to the west into the sodium bicarbonate zone which occupies a surface area of some 3 600 km^2. The third group occurs west of the second zone. It overlaps the area of Tulha-Adar-Fakhakhir where saline water is found in the Gezira Formation. Scattered patches of group 4 occur within either the sodium bicarbonate zone or the sodium chloride zone.

It appears that there is a progressive chemical modification in groundwater of the Gezira Formation. This modification coincides with the movement of water away from the Blue Nile. The hydrochemical evolutionary trend is from calcium bicarbonate water near the recharge zone of the Blue Nile to sodium bicarbonate resulting from base exchange and finally to a sodium chloride zone which coincides with the area of stagnation.

5.4 REFERENCES

Abdel Salam, Y. 1966, The ground water geology of the Gezira. M.Sc. thesis, Univ. Khartoum.

Abdel Salam, Y. 1976, The hydrochemistry of the Gezira Formation, Sudan. *Proceedings of the 3rd Conference on African Geology, Khartoum, January 1976* (in press).

Andrews, C.W. 1912, Notes on the molar tooth of an elephant from the Nile near Khartoum. *Geol. Mag.* 49:110-13.

Andrew, G. 1948. The geology of the Sudan. In J.D. Tothill (ed.), *Agriculture in the Sudan.* Oxford University Press:84-128.

Arkell, A.J. 1949, *The old stone age in the Anglo-Egyptian Sudan.* Sudan Antiquities Serv. Pamphlet 1:1-52.

Ball, J. 1939, *Contribution to the geography of Egypt.* Cairo:308.

Berry, L. 1964, Some notes on the relationship between geomorphology and soils in the Central Sudan. *Proc. Int. Geogr. Union Cong.* Liverpool.

De Wiest, R.J.M. 1965, *Geohydrology.* New York, John Wiley & Sons:366.

El Boushi, I.M. 1972, The shallow ground water of the Gezira Formation at Khartoum and the northern Gezira. *Sudan Notes and Records* 53:154-163.

Jewitt, T.N. 1955, Gezira soil. *Sudan Govt., Min. Agric. Bull.* 12.

Kheiralla, M.K. 1966, A study of the Nubian sandstone formation of the Nile valley. M.Sc. thesis, Univ. Khartoum.

Lacaille, A.D. 1951, The stone industry of Singa, Abu Hugar. The Pleistocene Fauna of two Blue Nile sites. *Fossil mammals of Africa 2.* London, British Museum (Natural History).

Lawson, A.C. 1927, The Valley of the Nile. Univ. California Chronicle 29.

Omer el Badri 1972, Sediment transport and deposition in the Blue Nile. M.Sc. thesis, Univ. Khartoum.

Ruxton, B.P. 1956, An account of the geology of the Sennar and Gedaref area. Unpublished manuscript, Univ. Khartoum, Geol. Dept.

Shukri, N.M. 1959, The mineralogy of some Nile sediments. *Quart. J. Geol. Soc. Lond.* 105:511-34.

Swan, C.H. 1956, *The recorded behaviour of the River Gash.* Min. of Irrigat. and Hydroel. Power, Khartoum.

Tothill, J.D. 1946. The origin of the Sudan Gezira clay plain. *Sudan Notes and Records* 27:153-83.

Tothill, J.D. 1948, A note on the origin of the soils of Sudan. In J.D. Tothill (ed.), *Agriculture in the Sudan.* Oxford University Press:129-143.

Whiteman, A.J. 1971, *The Geology of the Sudan Republic.* Oxford University Press.

Williams, M.A.J. 1966, Age of alluvial clays in the Western Gezira, Republic of the Sudan. *Nature* 211:270-1.

Williams, M.A.J. 1968a, A dune catena on the clay plains of the west central Gezira, Republic of the Sudan. *J. Soil Sci.* 19:367-378.

Worrall, G.A. 1956, A pedological study of Khartoum district. M.Sc. thesis, Univ, Khartoum.

R.H.GUNN
CSIRO, Canberra, Australia

6

The plains of the central and southern Sudan

6.1 LANDFORMS

The area considered in this account extends southwards from the Sennar to Kosti railway line to about the latitude of Roseires (12°N) on the Blue Nile and includes a narrow belt of land about 50 km wide bordering the White Nile between Kosti and Malakal. It is about 50 000 km^2 in extent and comprises two separate and distinct regions which are referred to broadly as the Blue Nile basin and the White Nile trough. Most of the area consists of plains which are mantled by dark alkaline clays. Narrow belts of recent alluvium border the Blue Nile and White Nile and there is an extensive area of undulating to rolling terrain underlain by windblown sand between Kosti and Keri Kera west of the White Nile. Scattered rocky hills or 'jebels', mainly gneisses, schists, marbles, and granites protrude through the mantle of superficial deposits.

The clay plains are similar to the Gezira, the triangular area to the north between the two Niles, where about 820 000 ha or two million acres of highly productive land is irrigated by gravity flow from the Sennar reservoir. A further 0.2 million ha are irrigated by pumping from the rivers. In addition, it has been estimated (Gibb 1954) that 500 000 ha could be irrigated by gravity flow from the Roseires reservoir and 226 000 ha by pumping from the rivers. In the more southerly parts of the regions where precipitation is adequate, extensive areas are suitable, or could be made suitable, for the production of raingrown crops.

The plains of the central and southern Sudan clearly, therefore, have very large potentialities for agricultural production and may be compared in many respects with the 'doabs' of West Pakistan and India and the Riverina region of southeastern Australia. They also border the middle reaches of one of the world's great rivers, the Nile. Considering the probable future importance of the regions, comparatively little effort has been given to the study of their biophysical features apart from the hydrology of the rivers.

The work and ideas reported here are based on 16 months' field work carried out on behalf of Sir Alexander Gibb and Parters in 1951-53 and on the survey data published in their report to the Sudan Government in 1954,

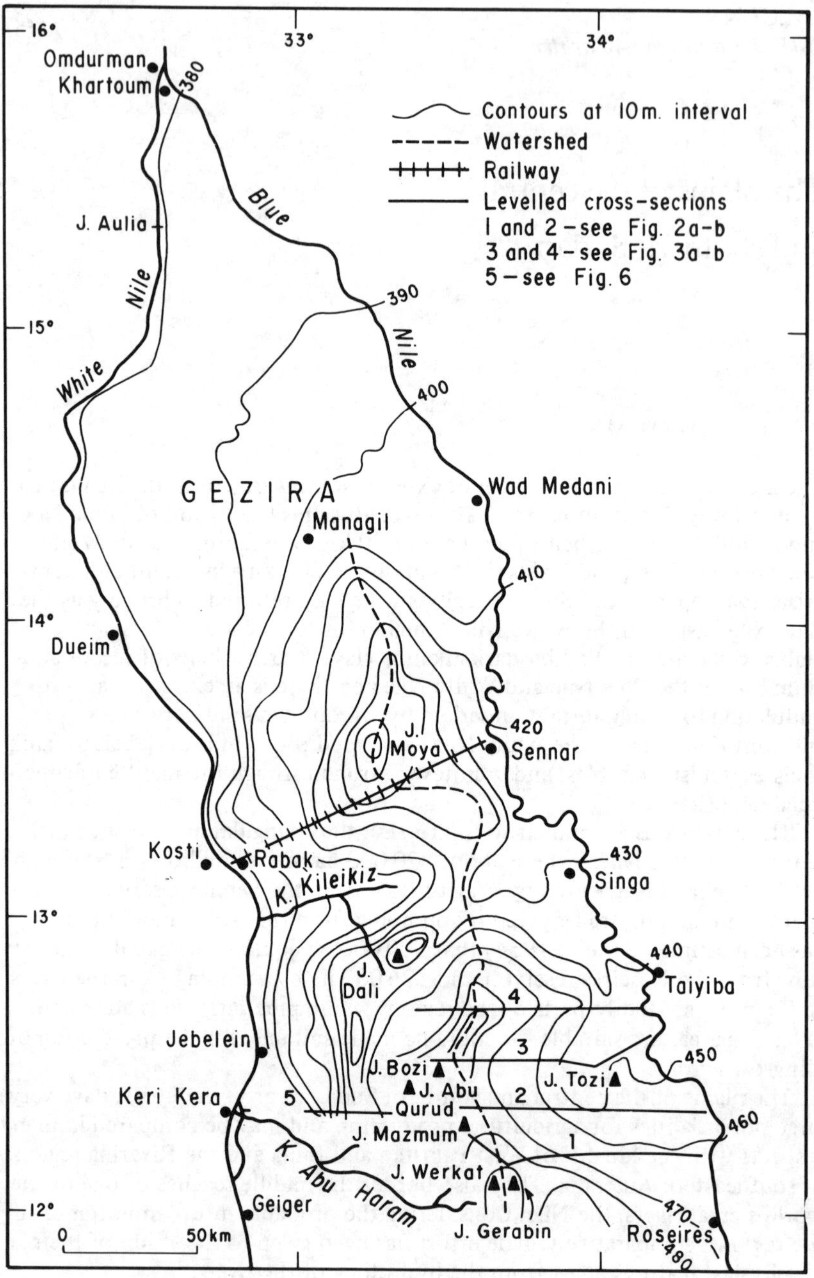

Figure 6.1. Contours between the Blue and White Niles.

which they kindly made available to me. The ERTS-1 imagery supplied by the United States Geological Survey has been of great value in providing a broad overall picture of the region. Much detailed work requires to be done, but this account of the landforms and soils is offered as a contribution towards a fuller understanding of the complexities of the Nile basin in the Sudan.

6.1.1 *The Blue Nile basin*

The Blue Nile rising in Lake Tana at an elevation of 1 840 m drains a catchment of about 200 000 km^2 in northern Ethiopia. Flowing south from the lake the river takes a large semi-circular bend and then flows in a northwesterly direction through a deep gorge up to 2 km deep and at Famaka, near the Sudan border, it has dropped to an elevation of 480 m (levels given here are based on the zero of the Khartoum gauge, assumed as 360 m above msl at Alexandria). The present gradients in the river decrease from 1:150 immediately below Lake Tana, to 1:1 000 near the Sudan border to about 1:10 000 below Sennar. Downstream from Famaka old meander patterns

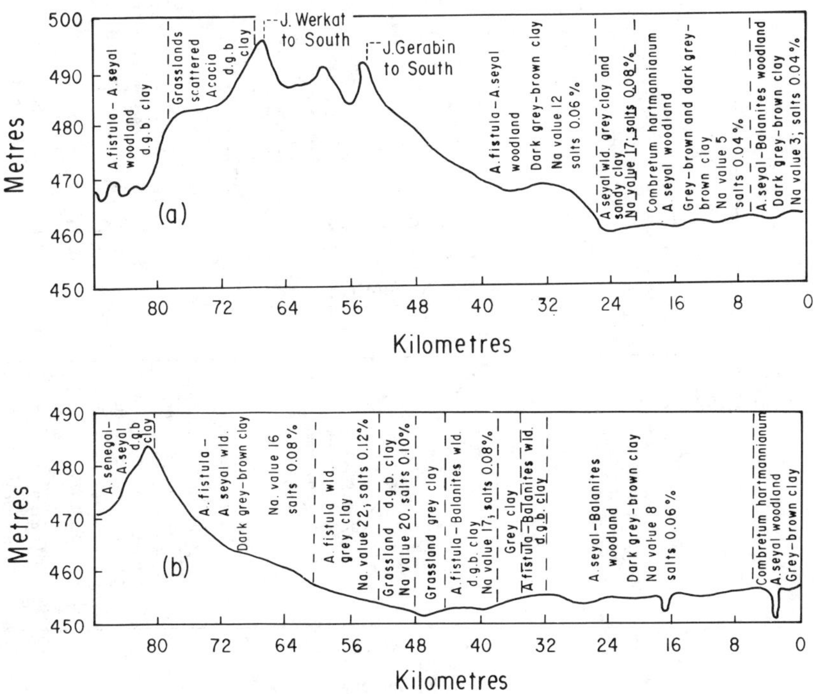

Figure 6.2. Cross-sections west of the Blue Nile:
a) latitude 12°10′N,
b) latitude 12°20′N.

visible in aerial photographs indicate that, before the river became deeply incised in its former deposits, gradients were probably lower than at present and the channel was subject to lateral shifts. Clearly, near the border there is a change from erosional to predominantly depositional processes where the river debouches onto the downwarped erosion surface on Basement Complex rocks and Nubian Sandstone in the Sudan (Andrew 1948).

6.1.1.1 *Landforms.* The clay plains on the interfluve between the Blue and White Nile rivers seem at first sight to be almost featureless, apart from the scattered jebels and changes in vegetation from north to south. Available contour maps and cross-sections west of the Blue Nile (Sudan Government 1929; Gibb 1954), however, indicate that there are several well-defined features:

1. Between Sennar and Roseires, there are two very gently sloping basins

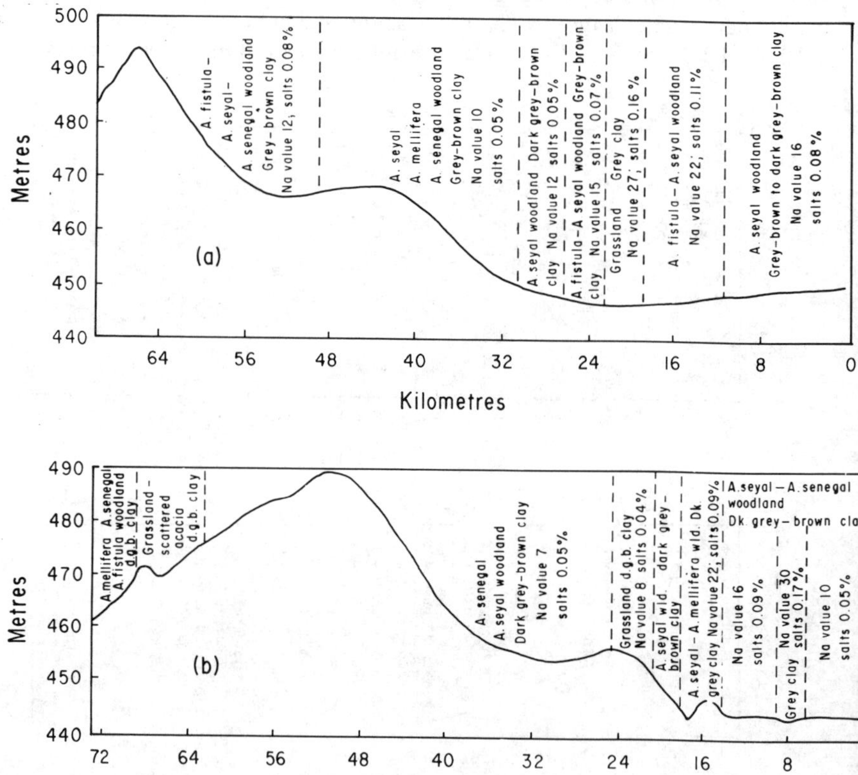

Figure 6.3. Cross-sections west of the Blue Nile:
a) latitude 12°30′N,
b) latitude 12°39′N.

extending westwards for up to 60 km from the river (Figure 1). These basins slope northeastwards at gradients of about 1:2 500 and are separated by a broad low ridge trending east-southeast from Jebel Dali to Jebel Ubeiyad. West of the southern basin the land slopes at a gradient of about 1:300 up to the watershed at an elevation of 490 m. To the west of the northern basin a low saddle in the watershed is only about 5 m above the general level of the plain to the east (Figure 1).

Four cross-sections obtained by levelling along lines at approximately 10 minute intervals for distances of up to 88 km west of the Blue Nile between latitude 12°10′ and 12°39′N (Gibb 1954) are shown in Figures 2 and 3. These sections show the almost level plains extending from 20 to 60 km west of the river with broad depressions and occasional relict channels 2 to 5 m deep with adjacent levees. On three sections (Figures 2a, 3a and 3b) there are quite abrupt changes in slopes up to broad terrace-like features with wide shallow depressions to the west and further abrupt slopes up to the watershed.

2. Between Abu Hugar and Sennar there is a belt about 20 km wide west of the Blue Nile with a series of old cut-off meanders and ox-bows indicating that the river has migrated eastward (Figure 4). West of the old meander belt is a broad ridge rising to 450 m which extends southeast from Jebel Moya (Figure 1). The ridge forms part of a belt extending from the latitude of Singa to north of Sennar which has a distinctly lighter tone on ERTS imagery than the adjacent land. The pattern may indicate a sandy surface cover and/or bedrock at shallow depth. Three very light-toned areas within the belt have similar patterns to those on the pediments of some of the larger jebels and in an area to the south of Rabak near the White Nile where ferruginized Nubian Sandstone is exposed.

3. The watershed and western plains and slopes. The low watershed between the Blue and White Nile rivers follows an almost direct northerly alignment from near Jebel Werkat to the east of Jebels Mazmum, Bozi, Abu Rawwag, and Dali at elevations of between 470 and 490 m (Figures 1 and 5). It then crosses a low saddle west of Singa at an elevation of about 435 m and rises to the ridge southeast of J. Moya at an elevation of 450 m. The longitudinal section is shown in relation to the approximate flood level of the Blue Nile in Figure 5.

The plains immediately west of the watershed have very gentle slopes of about 1:750 or less but gradients increase to between 1:300 and 1:400 farther west near the edge of a broad terrace east of the White Nile. The plains are broken by numerous jebels rising up to several hundred metres above the surrounding terrain, the largest area being occupied by the Jebel Abu Qurud-Abu Ureif-Terru complex believed to consist mainly of quartzite and graphite schist (J.R. Vail, personal communication) east-southeast of Jebelein. Numerous small khors and drainage lines flow south from this area to Khor Abu Haram which joins the White Nile 25 km south of Jebelein. Khor Kiliekiz originates south of Jebel Dali and reaches the White Nile about 23 km south of Rabak.

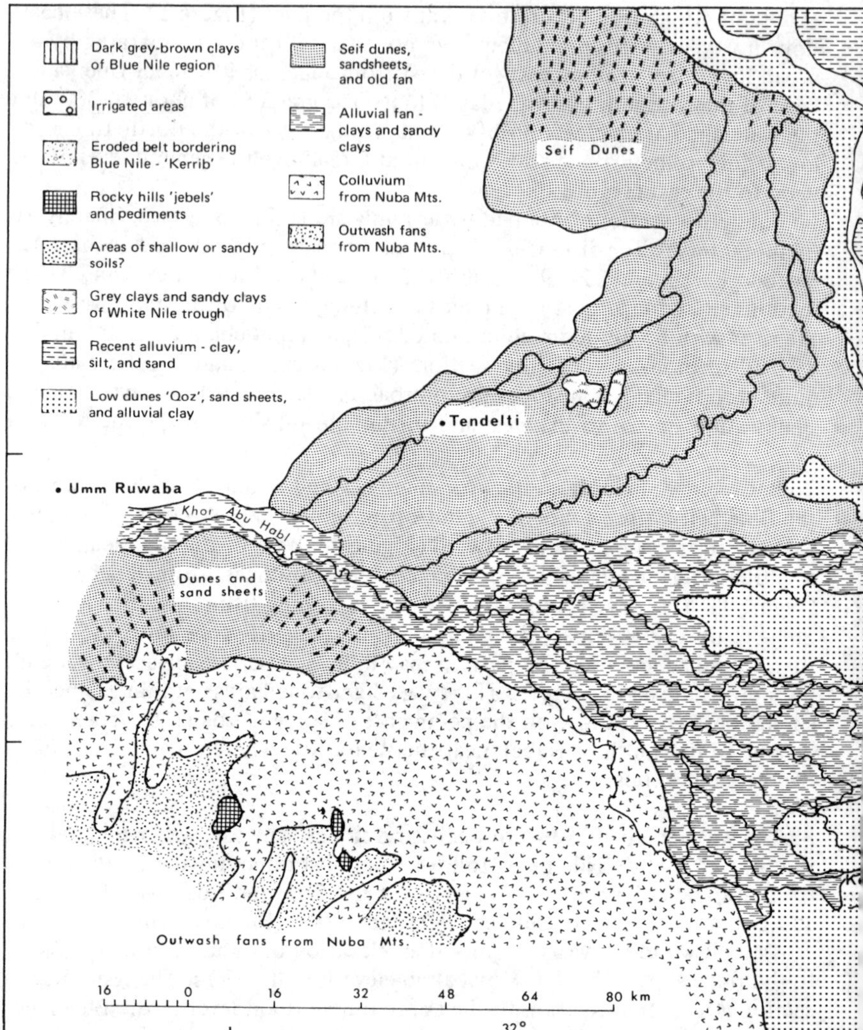

Figure 6.4. The Blue Nile clay plain, northern White Nile trough and adjacent landscapes, (Sketch map prepared from ERTS imagery taken in October and November 1972).

Apart from the reddish brown sandy loam soils on the pediments surrounding the jebels, all of this area west of the watershed is mantled almost entirely with dark brown to dark grey-brown clay. The western boundary is the break of slope at about 386 m which occurs at varying distances from the White Nile between Rabak and Gelhak where abrupt changes in gradient and soil occur. These features are shown on a cross-section east of the White Nile approximately in latitude 12°22′N (Figure 6). Here, the abrupt change in

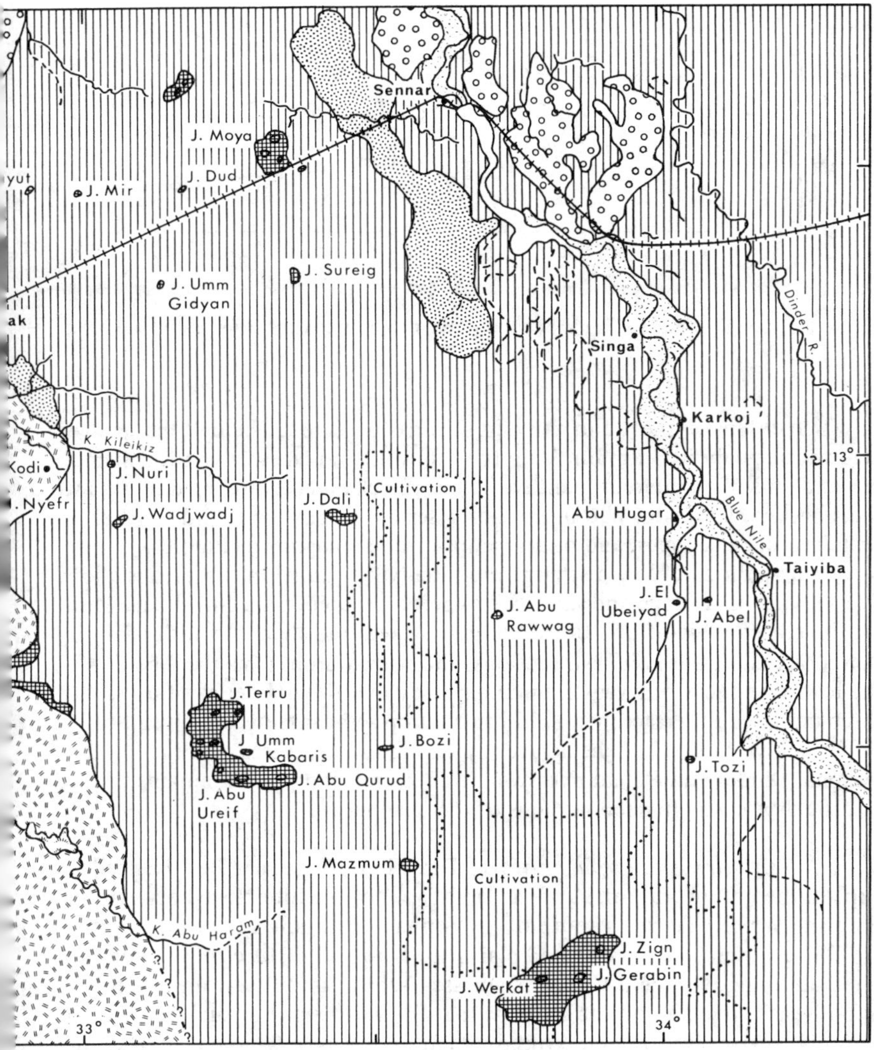

slope occurs 26 km east of the White Nile where the gradient changes from 1:6 500 on the White Nile terrace to 1:300 immediately east of this position.

4. The riverain features. Apart from rock bars at Damazin south of Roseires, and at Sennar, where dams have been constructed, the Blue Nile follows a meandering course which has become intrenched 20 to 30 m in the alluvium of the plain. On either bank there is an irregular belt known as the 'Kerrib',

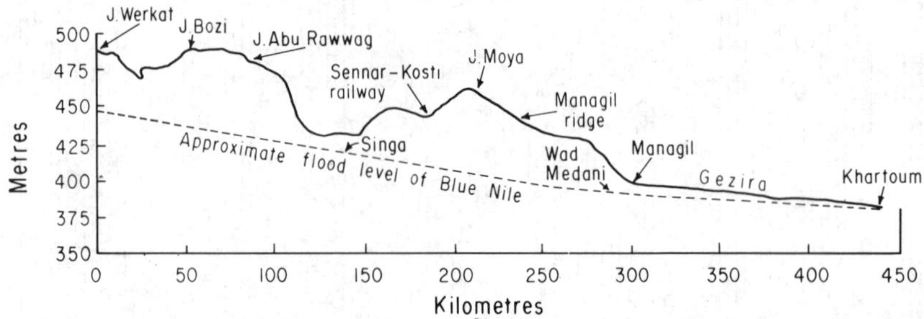

Figure 6.5. Longitudinal section along the watershed between the Blue and White Niles.

in which extensive sheet and gully erosion has occurred. The belt has an average width of about 3 km but is up to 4 km wide in places. It is not continuous and on some undercut, concave curves the channel is cutting directly into the adjacent clay plain. Within this belt a series of point bars, and terraces with levees, backslopes, and backswamps can be recognized (Figure 7).

During periods of low discharge the Blue Nile channel in this area is about 275 m wide on average and sands are exposed on point bars and shoals. During the seasonal high flows the sands are covered, the channel width is more than doubled in the point bar sections, and backswamps ('maiyas') behind the lowest terraces or levees are inundated by overflow. A series of up to three unmatched terraces with old backswamp remnants occurs on slip-off slopes below the level of the adjacent plain. Similar features occur along the seasonal Dinder and Rahad rivers where a series of long pools remains throughout the dry season in deep sections of the channels.

6.1.1.2 *Tentative geomorphic history.* Three hypotheses can be advanced to explain the nature and origin of the landscapes of the interfluve between the Blue and White Nile rivers south of the Sennar-Kosti railway line:

1. The watershed and elevated plains are erosional features developed on the underlying Basement Complex gneisses, schists, and limestones. This hypothesis is rejected on the grounds that the weathering of such rocks seems unlikely to give rise to fine-textured parent materials in which deep, alkaline clay soils could form. The soils on the pediments of the jebels which are scattered throughout the plains are reddish brown, massive sandy loams with neutral reaction. The formation of very deep sedentary soils would also imply a humid to sub-humid climate in this now semi-arid region at some stage during the Cainozoic.

2. A more feasible explanation would be that basic lavas were extruded over the region during the Tertiary volcanic episodes on the Ethiopian Plateau. The Gedaref ridge, forming the watershed between the Rahad and Atbara Rivers, was formed by sills which intruded the Nubian Formation and the soils in this area are similar superficially to those of the Blue Nile region (Andrew 1948). However, the sedentary soils of the Gedaref ridge are

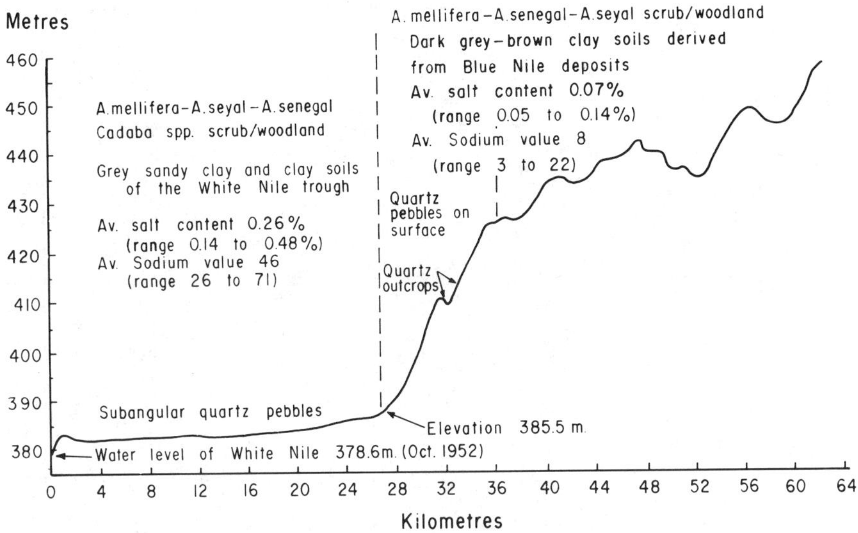

Figure 6.6. Cross-section east of White Nile in latitude 12°22′N.

only moderately deep (1.5 to 2.0 m), and they are underlain directly by fresh basalt and other rocks (B.P. Ruxton, personal communication). Similar soils occur in the extensive basalt region of the sub-humid Central Highlands of Queensland in Australia where stony knolls and ridges are bordered by areas of moderately deep, dark cracking clay soils underlain by basalt (Gunn 1974). Such soils have not been observed in the region under study and those that do occur are generally several metres deep and are underlain by alluvial sand near the rivers.

3. The hypothesis proposed here as being the most acceptable is that a large piedmont alluvial plain was built up by the Blue Nile, and its tributaries the Dinder and the Rahad. The Blue Nile is thought to have had a much greater discharge and a sediment load consisting almost entirely of clay, and to have built up a large alluvial fan over 50 m thick in places which extended from near the Sudan border to at least as far north as Managil. This hypothesis is expanded in the following section.

6.1.2 *The piedmont alluvial plains of the Blue Nile, Dinder and Rahad*

Uplift to 2-3 000 m asl of the Mesozoic sediments on the Ethiopian Plateau occurred in several stages during the Cainozoic and was followed closely by the extrusion of the Trap Series basalts which covered most, if not the whole of northern Ethiopia (Blandford 1869; Mohr 1971). The basalt-covered plateau was tilted slightly to the west (Grabham & Black 1925) and this tilting was the initial cause of the deposition of vast quantities of alluvium in Egypt and the Sudan (Nilsson 1953). Blandford (1869) suggested that at least

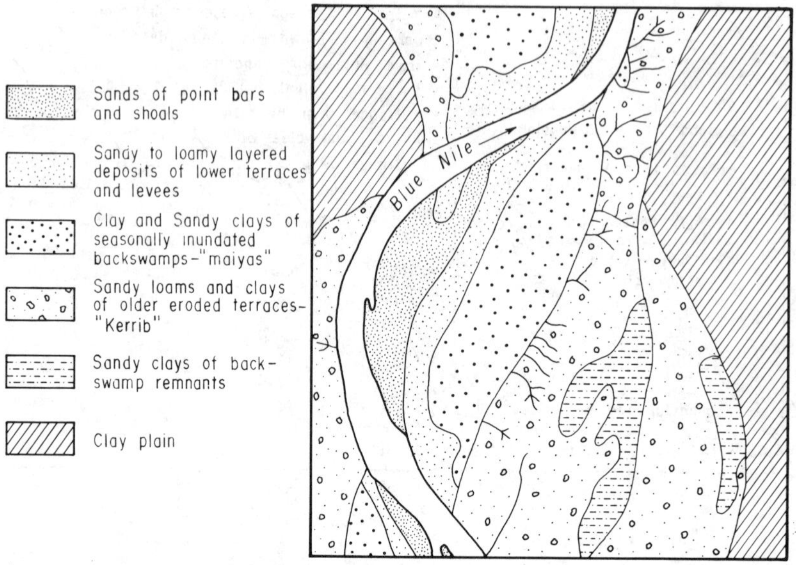

Figure 6.7. Riverain features of the Blue Nile c. 2 km south of Taiyiba, based on 1:25 000 air photograph.

1 200 m of basalt were removed. Assuming a thickness of 1 200 m and an age of 25 to 30 million years for the basalt, considered to be Oligocene-Miocene age by Mohr (1971), the calculated rate of erosion would be of the order of 45 m per million years.

At latitude 12°20′N the Blue Nile flows at an elevation of 60 m above the level of the White Nile, with a low divide separating the two basins. Judging from the nature of the deposits on and to the west of the watershed, it seems possible that during the formation of its alluvial fan the Blue Nile spilled over into the White Nile basin. Examination of the soils to depths of about 2 m at several hundred locations shows that the materials deposited were predominantly clays, and there is no evidence of sorting from south to north or of relict channels with sandy bedloads. From this evidence it may be concluded that the fan material was derived almost entirely from basalt before the river cut deeply into the underlying gneisses and quartzose sandstones from which the present bed load of sand is derived. That the suspended sediment was derived from fresh, not deeply weathered, basalt is indicated by the presence of readily weatherable minerals in the Blue Nile clays of the region (Shukri 1949) and by their obviously high content of swelling, montmorillonite clay.

Some information is available on the depth of the clay and the nature of the underlying materials. Near Jebel Tozi, about 13 km west of the river, Bunting & Lea (1962) recorded 5 m of clay and 1 m of sandy clay which were underlain by yellowish brown, bedded quartz sand with root-like, calcrete structures. A similar profile was seen by the author in the bank of the

Dinder south of Abu Hashim. Near Singa and Abu Hugar (Figure 4) Whiteman (1971) recorded 5 to 8 m of silty to sandy clay which was underlain by kankar gravel. In the Gezira, the depth of clay varies from 3 to 5 m at Shekeiniba (14°10′N, 32°02′E) near Managil to 110 m at Wad Adam (14° 21′N, 32°37′E) 50 km to the NW (Abdel Salam 1966). At Ghubshan (14° 14′N, 33°02′E), the type locality for the Gezira Clay, 17 m of dark clay and silt are underlain by 53 m of gravelly to clayey sand (Abdel Salam 1966; Whiteman 1971). A soil pit excavated to a depth of 4.3 m near Jebel Bozi on the watershed between the Blue and White Niles exposed a uniformly heavy clay throughout. In the traverse east of the White Nile at latitude 12° 22′N (Figure 6) vein quartz crops out at the surface and was encountered in two soil pits on the steeper slopes. Clearly, the depth of the clay would depend on the nature of the surface underlying the alluvial mantle. The scattered jebels and outcrops which protrude through the mantle and the longitudinal profile along the watershed (Figure 5) indicate that the surface was irregular. Abrupt changes in slopes occur on the western and northern extremities of the fan, some of which may be related to aggradation and degradation by the White Nile (see chapters 7 and 9).

Most of the jebels which protrude through the mantle occur near, or to the west of, the present watershed and indicate that a rocky ridge existed prior to the deposition of the clay. It is postulated that the river first built up a fan or lobe somewhere in the region of its present channel and then to the west, covering the rocky ridge with a mantle of clay and spilling over into the White Nile trough. Given the fine-grained sediments carried, it seems likely that the lobe was built up by a system of low-gradient, meandering, swampy distributary channels. The pre-existing relief was subsequently partly exhumed by lateral planation and general lowering of the surface by subsidiary drainage. Following a change in flow regime, consequent upon climatic change and/or tectonic uplift of the Ethiopian Plateau, there was a decrease in sediment load, and/or a change in the ratio of suspended load to bed load in the Blue Nile (Butzer & Hansen 1968). Vertical incision by the Blue Nile was accompanied by lateral planation of the eastern part of the fan as the main channel migrated in that direction. Lateral planation and downcutting by the Blue Nile are suggested by the wide terrace-like features evident on cross-sections between the river and the watershed (Figures 2 and 3) and by the broad depressions with grey clays which may indicate relict floodplains. Lateral streams originated near the scattered jebels where run-off rates were high and flowed off the watershed in both directions, dissecting the lobe to its present irregular form, and obliterating most of the evidence of prior river channels. Khor Doleib, flowing from the rocky Ingessana Hills, also lowered the landscape and deposited sand near the shoreline or upper terrace east of the White Nile near Renk.

No information is available on the age of the deposits on the watershed, but they are probably much older than the clays on the less-elevated plain west of the Blue Nile and those on the surface of the Gezira which were laid down in uppermost Quaternary times (Williams & Adamson 1973, 1974).

The uniformity and probably great depth of the deposits on the interfluve between the Blue and White Nile rivers and the lack of evidence of stratification or sorting indicate a long period of uniform conditions with respect to climate, river regime, source materials, and depositional environment. Further, the great volume of material transported suggests prolonged erosion in the catchment. According to Butzer & Hansen (1968), the primary source area of Nile sediment is the sub-tropical Voina Dega region at 1 800-2 700 m in Ethiopia, where vertisols rich in montmorillonite clay predominate. This is confirmed by the cracking, self-mulching characteristics of the soils on the plains in the Sudan and by one analysis from near Sennar (El Attar & Jackson 1973). At higher elevations in the temperate Dega region, red clay soils with dominantly kaolinite or halloysite minerals occur. It is possible that a deep weathering mantle covered the basaltic regions during the Tertiary and this was later stripped off to expose the underlying fresh rock following a change to a drier climate and/or renewal of tectonic uplift. An analogous situation occurred in central Queensland (Gunn 1974; Gunn & Galloway, 1978) where the Quaternary was characterized by alternating periods wetter and drier than the present. Such climatic fluctuations would account to some extent for the widespread erosion in Ethiopia, as well as alluviation in the Sudan which may have been correlated with periods of waning river discharges (Butzer & Hansen 1968). If such low runoff conditions prevailed at the high elevations in Ethiopia, the climate on the Sudan plains would have been arid to semi-arid and probably somewhat drier than at present. The presence of abundant calcium and magnesium carbonate in the parent materials of the soils on the watershed near Jebel Bozi and elsewhere supports such a view (Flint 1959).

This interpretation of the origin and mode of formation of the landscapes on the interfluve between the Blue and White Niles may appear somewhat tenuous, but it is offered as a working hypothesis, with a realization that much work is necessary to provide definite evidence. Drilling across the interfluve to determine the thickness of the clay mantle and the nature of the underlying materials is one suggested line that might be followed. A study of the mineralogy of samples collected from the bores would assist greatly in determining the provenance of the sediments and an investigation of the materials exposed in Khor Dunya, west of the Roseires reservoir, and in other sections further upstream may provide evidence on the probable elevation and southern limits of the clay mantle.

6.1.3 *The White Nile trough*

The White Nile north of Malakal collects drainage from the western Ethiopian Plateau via the Sobat, from the East African lakes region via the Bahr el Jebel, and from the Nile-Congo divide via the Bahr el Arab and Bahr el Ghazal. The Sobat which enters the White Nile near Malakal contributes 60 and 25 % of White Nile discharge during high and low levels respectively. About half the discharge of the Bahr el Jebel and most of the Bahr el Ghazal flow are lost by evapotranspiration in the extensive permanent swamps of the Sudd (Hurst 1952). Prior to about 12 000 years ago, when there was no outlet from Lake Victoria (Kendall 1969), the Bahr el Jebel was probably no more than a local seasonal stream but about 9 500 and 6 000 years ago the lake was particularly high (Butzer et al. 1972) and it was probably during these periods that the great swamps of the Sudd were formed (Williams & Adamson 1973).

The average gradient of the river between Malakal and Kosti is 1.4 cm/km or 1:70 000 and, according to Andrew (1948), the profile is that of an internal drainage basin aggraded to the outlet level. The base level of the White Nile prior to the construction of the Jebel Aulia dam was controlled by the discharge and depth of channel incision of the Blue Nile at Khartoum. It is feasible that when the Blue Nile flowed northwestwards across the Gezira (Williams & Adamson 1973) it would have retarded or prevented outflow from the White Nile trough where a lake or seasonal swamp would have formed. Evidence has been recorded of former terraces or shorelines east of the river at about 380 ± 2 m (Tothill 1948) and 386 m (Gibb 1954; Berry 1961). The collection of sub-fossil freshwater mollusca east of the White Nile north of Rabak and C^{14} dating by Williams (1966) and Williams & Adamson (1973, 1974) confirm that between 8 000 and 12 000 years ago the White Nile was at least 20 km wide and up to 5 m higher than the normal flood level before the construction of Jebel Aulia dam in 1937-42.

The sediments which have filled the basin are clearly derived from various sources, including basalt and other rocks on the Ethiopian highlands and Basement Complex and volcanic rocks in the East African catchment and the Nuba mountains. Streams flowing from the erosional landscapes on granite and gneiss in the Ingessana Hills, Jebelein, and other rocky areas on the watershed between the White and Blue Niles also contributed. If the concept of the Blue Nile fan is correct, then material from this source would also have been deposited in the northern part of the White Nile trough. According to Shukri (1949), the White Nile sediments contain sillimanite, garnet, staurolite, epidote, and kyanite and differ from those of the Blue Nile deposits which contain abundant hornblende, biotite, iron ores, augite, and lava fragments. However, as Williams & Adamson (1973) point out, the distinction between heavy mineral assemblages from the two sources is not clear-cut and much further work is necessary. Clay mineral analysis in respect of two surface soil samples from near Kosti and Melut indicates that montmorillonite is dominant and that vermiculite, kaolinite, amorphous material, and mica are present in minor proportions (El-Attar & Jackson 1973). Clearly also in this respect further work is required before an assessment of the clay mineralogy of the White Nile deposits can be made.

On the basis of differences in climate, topography, vegetation, landforms, and the nature and origin of soil parent materials, the White Nile trough between Kosti and Malakal can be divided into two distinct zones:

1. *Northern Zone between Kosti and Gelhak*

This zone includes the almost straight, north-flowing section of the river about 420 km long with several large islands. It is bordered on the east by a

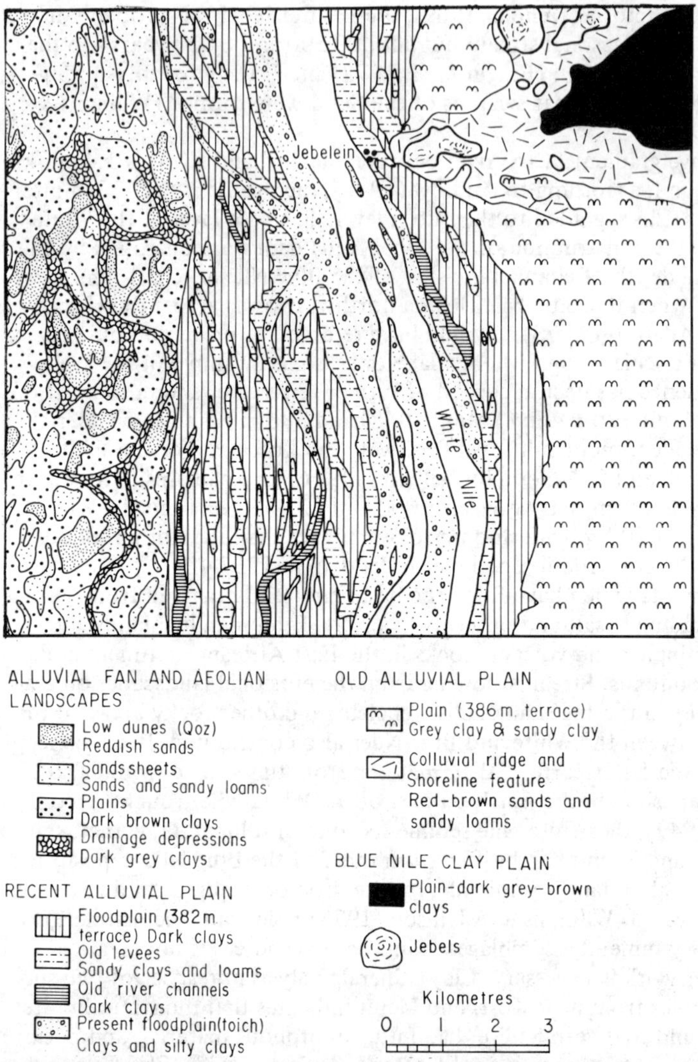

Figure 6.8. Landforms and soils near Jebelein on the White Nile.

strandline or marked change of slope at varying distances from the river which forms the boundary between the Blue and White Nile deposits (Gibb 1954). Clays and sandy clays occur on low and high terraces respectively and correspond approximately with 382 and 386 m levels in this area. A narrow flood plain (toich), which is seasonally inundated, occurs mainly to the west of the river channel.

Imagery from the ERTS-1 satellite taken in October 1972 clearly shows that on the left bank between Kosti and Keri Kera, a large alluvial fan with numerous small meandering distributaries, has been formed by Khor Abu Habl which drains the northern slopes of the Nuba Mountains (Figure 4). Near the river in this section there is a complex pattern of fixed dunes (qoz) and sandsheets interspersed with irregular areas of clay. This pattern ends abruptly at the edge of the belt of recent alluvium at a level of about 382 m. The source of the sand is the extensive spread of sandsheets and seif dunes to the northwest but reworking by wind action of sand from the beds of distributary channels in the Khor Abu Habl fan is also probable.

Landforms. Apart from a common belt of recent alluvium up to 6 km in width, the landforms to the east and west of the White Nile between Kosti, to about 30 km south of Keri Kera (13°10′N to 12°00′N) differ markedly owing to dissimilarities in the nature and mode of formation of the depositional materials (Figure 8).

At distances varying between 4 and 30 km east of the river between Rabak and Gelhak there is a very gently sloping terrace at levels of between about 382 to 386 m. At the eastern limit there is a ridge or an abrupt change to steeper slopes which marks the eastern border of the White Nile trough (Figures 4 and 6). This feature has been traced as far south as 11°00′N to a point about 30 km east of Gelhak and is clearly visible on ERTS imagery at 1:1 million scale. It is probable that it then follows a southeasterly direction to the south of the Ingessana Hills and then south along the footslopes of the Ethiopian highlands to form the bulge in which the Mabaan area and Machar Marshes occur. Except where a few large khors have cut shallow channels across the terrace, there is no evidence of the intricate pattern of levee remnants which is present on the recent alluvial terrace below 382 m (Figure 8). Just south of Rabak, Nubian sandstone is exposed in a low cliff, and near Jebelein the eastern border is marked by a ridge of reddish brown sand derived at least partly from the granite hills nearby, and possibly also from sandy deposits on the shoreline of a former lake. Massive ferruginous material crops out in this area and a sparse strew of rounded to sub-angular quartz pebbles occurs on the surface. Near Geiger and Renk there are similar sandy features which have probably formed on materials deposited by khors flowing from the rocky Ingessana Hills to the east. The feature near Renk is known as Qoz Ngyok where a catenary sequence of soils was described by Jones (1948). Hard compacted, reddish-brown sandy loams occur on the terrace edge about 30 km east of Gelhak.

On the left bank between Kosti and Keri Kera above the level of 382 m, the upper limit of the belt of recent alluvium, there is a complex pattern of dunes, sandsheets, and shallow discontinuous linear drainage depressions interspersed with irregular areas of dark clay. These features occur along the base of the large alluvial fan developed by Khor Abu Habl. The apex of the fan is 40 km east of Umm Ruwaba (Figure 4). The dunes are 10 to 15 m high and irregular in shape, generally with the longer axes aligned in a north to northeasterly direction. Most are stabilized by low woodland of *Acacia senegal* and a sparse cover of short annual grasses, but where the cover has been destroyed by clearing for cultivation they are subject to deflation. Sheets of

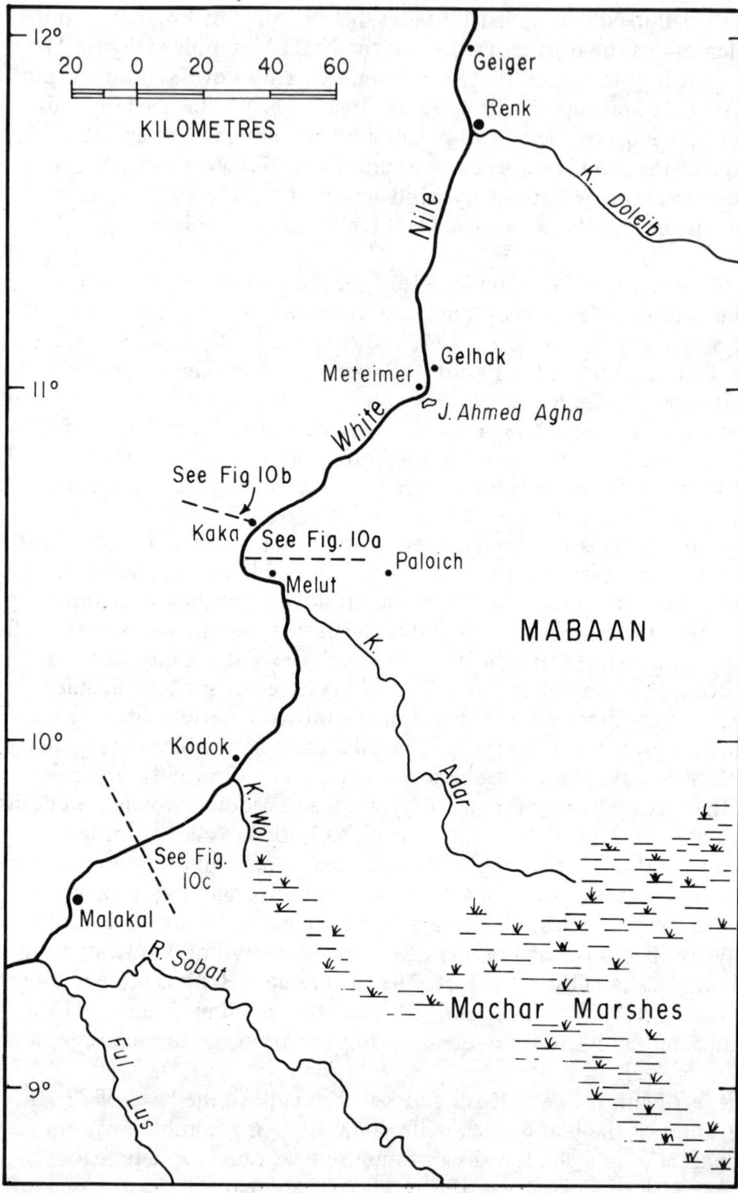

Figure 6.9. The White Nile southern zone.

sand with a cover of Acacia scrub commonly border the dunes or occur as irregular, isolated areas surrounded by alluvial clay. The linear drainage depressions have clay floors and are subject to seasonal flooding. They appear to be inactive and are probably relict distributary channels of the fan.

Irregular belts of recent alluvium occur on both sides of the White Nile channel which is about 1 km wide during low discharge. Assemblages of four well-defined landforms can be distinguished in this zone. Nearest the channel and rising to a level of about 378 m is the present floodplain or 'toich' up to 2 km wide with a cover of hydrophytic grasses, which is seasonally inundated to depths of between 40 to 165 cm for periods of 100 to 250 days in the Jebelein area (Basiński & Davies 1954). Subsidiary narrow channels and linear pools cut the floodplain in places and there are scattered low linear banks of sandy clay. Bordering the floodplain are low levees about 2 m high with sandy loam to sandy clay textures. They form almost continuous features but are broken in places by transverse drainage lines. The old floodplain of clay, which is underlain in places by pale yellow sand, generally has an uneven micro-relief, and levee remnants and ridges form low linear features 1 to 2 m high, aligned parallel to the relict channels. Runoff from the hard compacted ridge surfaces has formed shallow rills and gullies along the edges. Levels rise gradually from about 378 to 382 m where there is a clear boundary to the complex landscapes of the Khor Abu Habl fan (Figures 4 and 8). On 'debbas' or habitation sites on the oldest levee a few kilometres south of Keri Kera polished stone implements and sherds are scattered on the surface and layers of ash and charcoal were noted in a soil pit in this area.

The recent alluvial features described above may be equated with those east of the river north of Kosti, where subfossil shells of *Lanistes* and *Cleopatra* sp. were collected and dated at 8 000 to 12 000 years BP by Williams (1966). Evidence in support of this contention is the collection by the author in 1952 of sub-fossil shells (identification by staff of the Natural History Museum, Kensington, London) embedded in calcrete at a depth of 1.5 m in a pit 9 km north of Jebelein. The elevation at this site was approximately 380 m.

Four alluvial islands (Musran, Bulli, Renk & Ahmed Ahga) all about 38 km long occur in the northern zone and have been described by Berry (1962). That author concluded that the islands were formed in mid-channel when the river discharge was about 10 times greater than at present, although this estimate was questioned by Tricart in a postscript to the same paper. There is evidence in the form of old river courses west of the present channel, that other large islands existed at one time between latitudes 10°48′N and 11°30′N. The presence of a thick layer of pale yellow sand beneath about 1 m of gypseous clay in several soil pits near Meteimer in this area, appears to support the idea that river discharge and velocity were high enough to transport a bedload of sand, although an aeolian origin for some of the sand remains possible.

2. *Southern Zone between Gelhak and Malakal*

This zone comprises the predominantly northeast-flowing section of the river

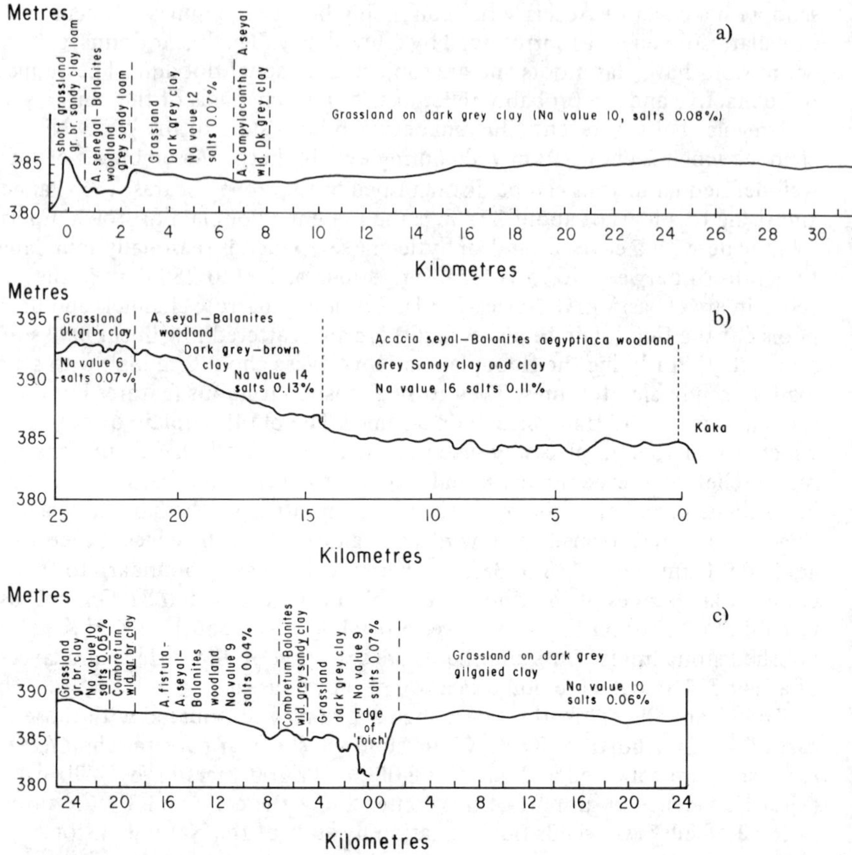

Figure 6.10. Cross sections in the White Nile trough: a) Melut bend, 10°30′N, b) west of Kaka, 10°35′N, c) latitude 9°40′N, east and west of White Nile.

about 260 km long with the prominent westward bend at Melut (Figure 9). The river is bordered by extensive clay plains which are generally poorly drained and covered mainly by perennial tussock grassland with scattered areas of open woodland. The eastern plain includes the Mabaan area and the Machar marshes which drain to the river via Khor Wol and Khor Adar and numerous smaller channels. A 32 km cross-section east of the river from near Melut to Paloich shows a prominent sandy levee rising to 385 m, a narrow back swamp, and an abrupt slope up to the almost flat clay plain at a level of 383 to 384 m (Figure 10a). A section west of the river from Kaka shows an almost flat profile at an average level of 385 m, a short abrupt slope at 386 m about 15 km west of the river, and a gradual rise to about 393 m at 25 km (Fig 10b). Two sections on either side of the river in latitude 9°40′N show almost leve plains with gilgaied microrelief at an elevation of 387.5 m (Fig. 10c). Hard compacted sandy levees, parallel to the river, together with grassy floodplains about

2 km wide broken by swampy secondary channels, are prominent features of this southern zone. The Sobat joins the White Nile about 16 km south of Malakal and the deposition of reddish-brown clays and silty clays along the northern bank indicate that the alluvium has been derived, at least in part, from red clays formed on deeply weathered basalt on the Ethiopian highlands and foothills (Morin & Parry 1971).

Apart from the three cross-sections shown in Figure 10, which extend for 25 to 30 km from the river, and a number of shorter sections, little factual information is available on the extensive plains of the southern zone. In the section east of Melut (Figure 10) the clay plain, broken by shallow drainage lines in places, is entirely below 385 m. A pit excavated to a depth of 7 m in this area exposed a profile entirely of heavy clay changing in colour from very dark grey in the upper 2-3 m to yellowish-brown with prominent mottling beneath, where carbonate concretions increased with depth. West of Kaka (Figure 10) there is an almost level plain at 385 m of grey clay and sandy clay extending for about 15 km from the river with a nick point at 386 m which indicates the possible existence of a terrace. The origin of the dark grey-brown clay plain rising gradually to 393 m between 15 and 25 km west of the river is not known. It may be speculated that this gently sloping area was formed in colluvial deposits from the highlands north of Talodi or, alternatively, that it was formed in an extensive lake rising to about 400 m as postulated by Ball (1939). With the exception of an area broken by old levees and channels below 386 m on the left bank, the section in Figure 10c shows a gentle rise to about 389 m while on the right bank the almost flat plain has a general level of 387.5 m. The ubiquitous remains of the amphibious mollusc *Pila wernei* on the right bank plain is evidence of seasonal flooding and waterlogging. The relationship of this land to the 386 m terrace further north is uncertain, but it is possible that recent aggradation may account for the increase in level.

6.2 SOILS

This account of the soils covers the regions referred to as the Blue Nile basin and the White Nile trough described in Part 1. It is based mainly on data collected during the irrigable areas survey by Gibb (1954) in which about 1 000 profiles were examined by means of pits excavated to a depth of 1.8 m. Analytical data on samples collected from these pits were provided by the Research Division of the Ministry of Agriculture, Sudan Government. The scope of the work reported here is essentially reconnaissance in nature and is based on 16 months in the field. Mapping of the soils, mainly in a narrow belt on both banks of the White Nile, was carried out by means of aerial photographs taken in 1952 at a scale of 1:25 000 (Gunn 1955). Further information on the soils of these regions is provided by the Kenana soil survey carried out by the Sudan Government in 1950-51 and the unpublished report on this work by T.N. Jewitt who kindly supplied a copy of his report. The soils and vegetation of the eastern half of the Blue Nile region were described by Bunting & Lea (1962) and relevant information on soils, mainly in respect

Table 6.1. Physical and chemical properties of soils on the Blue Nile Clay Plain. (Particle size range from clay to gravel).

Location	Soil colour and number of profiles	Particle size (%)					Soluble salts (%)	Na value me/100 g clay	pH 0-30 cm	pH 30-60 cm	pH 60-90 cm
		Gr	CS	FS	Si	C					
Lat. 12°10′N (Fig. 6.2a)	Grey-brown to dark grey-brown (40)	1	2	14	8	75	0.06	11	8.5	8.9	9.1
	Grey (2)	2	7	27	9	55	0.08	17	8.8	9.4	9.4
Lat. 12°20′N (Fig. 6.2b)	Grey-brown to dark grey-brown (32)	1	3	17	9	70	0.09	13	8.6	8.9	9.0
	Grey (5)	3	4	19	10	64	0.10	17	8.8	9.1	9.3
Lat. 12°30′N (Fig. 6.3a)	Grey-brown to dark grey-brown (30)	2	3	17	8	70	0.07	13	8.7	9.0	9.0
	Grey (5)	2	4	20	9	65	0.13	23	9.2	9.4	9.2
Lat. 12°39′N (Fig. 6.3b)	Grey-brown to dark grey-brown (34)	1	4	15	9	72	0.05	9	8.7	9.0	9.2
	Grey (3)	2	6	21	10	61	0.12	25	9.3	9.5	9.2

Table 6.2. Dark grey-brown clay soil on Blue Nile deposits: analytical data

Depth cm	Particle size distribution (%)					pH (1:5 soil/water)	Soluble salts (%)
	Gr	CS	FS	Si	C		
0-30	1	1	15	13	70	8.45	0.04
30-60	2	2	15	10	70	8.65	0.05
60-90	1	2	15	10	71	8.70	0.07
90-120	1	2	12	13	73	9.00	0.07
120-150	1	2	14	10	73	9.10	0.08
150-180	3	1	11	12	73	9.00	0.10
180-245	2	2	13	10	74	9.10	0.13
245-305	1	1	12	11	75	0.05	0.12
305-365	6	3	12	14	65	9.20	0.10
365-430+	4	1	13	10	72	9.00	0.11

of the Gezira area to the north, is recorded by Tothill (1948) and co-authors.

The dominant soils throughout the two regions are dark, alkaline cracking clays most of which may be classified as 'grumusols' in the terminology of Oakes & Thorp (1950) or 'vertisols' according to the 7th approximation system of the US Department of Agriculture (1960). There are however, significant variations in physical and chemical properties of the soils, both within and between the two regions, which are related to differences in the nature and mode of formation of the parent materials. Except for small areas of soils formed *in situ* or on colluvium near jebels, more extensive areas on aeolian materials, and the alluvium of the Khor Abu Habl fan west of the White Nile, these soils have formed in materials that have been transported and deposited by the Blue and White Nile rivers.

The analytical data provided by the Sudan Government are based on routine analyses of particle size distribution, soluble salts, and sodium value in bulked samples from the upper 90 cm and pH measurements at 30 cm intervals to 90 cm depth. The sodium value is the soluble plus exchangeable sodium expressed as milli-equivalents of sodium per 100 g of clay (Jewitt 1955), and cotton yields in the Gezira and elsewhere have been positively correlated with variations in this parameter.

6.2.1 *The Blue Nile basin*

The soils of the clay plains in the eastern part of this region were examined mainly along four traverses extending up to 88 km west of the Blue Nile with soil pits sited at 2 km intervals. The topography, soils, and vegetation along these traverses are shown in Figures 2 and 3 and analytical data in respect of the soils are summarized in Table 1. Away from the eroded belt bordering the river and the sedentary soils surrounding the jebels, the cracking clay soils of the plains convey a first impression of remarkable uniformity. There are, however, significant if subtle differences in soil colour, and in the self-mulching properties of surface horizons which are correlated with relatively minor variations in relief, not readily apparent on the ground.

The dominant colours of the surface soils are dark grey-brown to very dark grey-brown (10YR 4/2-3/2) but in shallow depressions 1 to 4 km wide, dark grey to grey colours (10YR 4/1-5/1) predominate which are associated with poor drainage and seasonal waterlogging. The general pattern of distribution of these soils is shown in Figures 2 and 3. Bunting & Lea (1962) described an assemblage of soils near Jebel Tozi and related colour differences to changes in micro-relief. They found grey-brown soils (2.5Y 5/2) in shallow closed depressions which are subject to prolonged seasonal flooding, dark grey-brown soils (2.5Y 4/2) at slightly higher elevations (up to 70 cm), and very dark grey-brown soils (10YR 3/2) in transitional areas.

Differences between soils are also apparent in the formation on drying of very friable, fine granular self-mulching layers 10 to 20 cm thick at the surface. Soils which form granular layers are most extensive on the watershed between the two Nile rivers in the vicinity of Jebel Bozi and are termed 'fuda' soils by the local cultivators. In other areas, particularly in seasonally flooded or waterlogged sites the soils do not form granular mulches but dry out to form hard, cracked surfaces. Possible reasons for the formation of thick self-mulching layers are high contents of montmorillonite clay and/or the presence of finely-divided carbonate.

Broad correlations between soil colours and some physical and chemical properties are indicated by the figures shown in Table 1. The grey soils have lower contents of clay and somewhat higher contents of fine sand than the dark grey-brown soils. Soluble salt contents, sodium values, and pH values in the grey soils are all invariably higher than in the dark grey-brown soils, presumably owing to the accumulation of soluble salts and exchangeable sodium in the low-lying situations. In respect of the grey soils particularly, the figures also show a tendency for the soluble salt contents to increase from south to north which probably reflects the decrease in rainfall northwards and the less intense leaching.

The examination of several hundred profiles in pits excavated to depths of 1.8 m showed very uniform colours and textures. Below about 1.8 m there is a gradual change to brown or strong brown colours and carbonate nodules and concretions become more abundant with depth. A soil profile exposed to a depth of 430 cm on the watershed between the Blue and White Niles near Jebel Bozi is described below and analytical data are given in Table 2. These figures show the very uniform textural properties and gradual increase in soluble salts and pH values with depth. The gravel fraction (>2 mm in diameter) comprises carbonate nodules and concretions. As suggested earlier, the characteristics of this profile provide some evidence in support of the hypothesis that a uniform clay mantle was deposited by the Blue Nile.

Location: 6 km north-east of Jebel Bozi, latitude 12°30′N, longitude 33°30′E.
Elevation: 490 m asl.
Average annual rainfall: 670 mm.
Vegetation: Acacia scrub/woodland – *A.mellifera* and *A.fistula* co-dominant; *A.senegal,* frequent; *A.seyal* and *Balanites aegyptiaca,* infrequent; scattered small shrubs – *Combretum aculeatum, Cephalocroton cordofanus, A.campylacantha.* Annual grasses – *Sehima ischaemoides, Cymbopogon nervatus,* and *Beckeropsis nubica.*

Depth (in cm)	
0-30	Very dark grey-brown (10YR 3/2) heavy clay; thin (1-2 cm) granular, strongly self-mulching layer, soft and friable; coarse prismatic structure beneath, breaking to medium angular blocky; firm to hard; a few small pale yellow carbonate nodules; many roots; large vertical cracks to 75 cm; gradual transition to:
30-60	Dark grey-brown (10YR 4/2) heavy clay; as above; moderate root development.
60-245	Dark grey-brown (10YR 4/2) heavy clay; coarse angular blocky structure, hard; slight increase in carbonate nodules at 180 cm; few roots to 150 cm; very gradual transition to:
245-430+	Brown (10YR 5/3) heavy clay; coarse angular blocky, hard; abun-

dant large pale yellow carbonate concretions; no roots; some termite holes and fungus gardens.

In the northern part of the Blue Nile region, the Sudan Government carried out soil surveys in 1950-51 and collected samples from 500 soil profiles. T.N. Jewitt (unpublished report) provided summaries of the analytical data available for these soils. The survey covered the area between the two Niles extending southwards from the Sennar-Kosti railway to latitude 12°45′N and the soil pits were located in three compact groups referred to as the eastern, central and western Kenana areas. The western area extended on to the 386 m terrace east of the White Nile where the soils are markedly different from those immediately to the east on Blue Nile deposits. No profile descriptions are available from these areas, but the following information on chemical properties is based on Jewitt's report.

In the eastern area the soil profiles examined are located to the west of the belt of old cut-off meanders and ox-bows north of Abu Hugar which is shown in Figure 4, and in places extended into the belt where dark grey clay and sandy clay soils occur in broad shallow depressions. Some of the chemical properties of three groups of profiles in this area are given below:

	Range in salt content %	Average sodium value	
75 profiles	0.10 - 0.20	21	
29 profiles	0.21 - 0.30	24	range 3-39
17 profiles	0.31 - 0.40	25	

The 17 profiles with the highest salt contents are almost certainly related to the dark grey soils in the depressions of old backswamps and prior channels.

The central area extends from just north of J. Sureig southwards to J. Dali (Figure 4) where grey-brown soils with remarkably uniform physical and chemical properties occur. The mean sodium value and soluble salt content are 12 and 0.14 % respectively.

The western area is west of the J. Umm Gidyan - J. Dali road and extends southwards to J. Nyefr near the White Nile. Excluding the grey soils of the White Nile trough, the grey-brown soils are very similar to those in the central area and the following data are available for two groups of profiles:

	Average salt content %	Sodium value
64 profiles	0.15 (range 0.10 - 0.20)	11
29 profiles	0.24 (range 0.21 - 0.30)	14

6.2.1.1 *Comparison of the clay soils of the Kenana with those of the Gezira.* Descriptions of the soils that occur in the irrigated areas of the Gezira indicate that they differ in many respects from the soils of the Kenana district south of the Sennar-Kosti railway. Earlier in this chapter, attention was drawn to the topographic differences between the two regions (see Figures 1 and 5) and to the probable younger ages of the soils in the Gezira as well as in the two basins west of the Blue Nile farther south, compared with those on the more elevated plains.

The typical Gezira soil is a self-mulching, strongly alkaline, dominantly montmorillonitic, cracking clay (50-65 % clay) and is underlain by sand and gravel at depths of 3 to 4 m or more. The principal physical and chemical properties are shown in Figure 11. Some areas of this soil have been irrigated for 50 years and Jewitt & Middleton (1955) concluded from their investigations that there is no evidence of salts rising or accumulating in the soils and that there is in fact, some evidence of salt removal from profiles to a depth of 180 cm. The occurrence of a moderately saline, flocculated subsoil together with the use of high quality irrigation water, a high application efficiency, and the growing of salt-tolerant crops are factors which have contributed towards the successful, long-term use of these soils.

The profiles of the dark grey-brown soils of the Kenana are more uniform and there is no grey subsurface horizon. Average clay contents also appear to be somewhat higher (see Tables 1 and 2). The salt contents are appreciably lower than those shown in Figure 11 and the soils generally do not contain gypsum. Jewitt (unpublished report) suggests that flocculation in the Gezira clay occurs where the salt content rises to about 0.25 % or higher, whereas in the Kenana soils the salt content rarely reaches this level. These differences in the physical and chemical properties between the soils are almost certainly related to differences in the nature and mode of formation of the parent materials and also, possibly, to age and climate.

6.2.2 *White Nile trough*

We have seen that the alluvial and aeolian landscapes in the White Nile trough

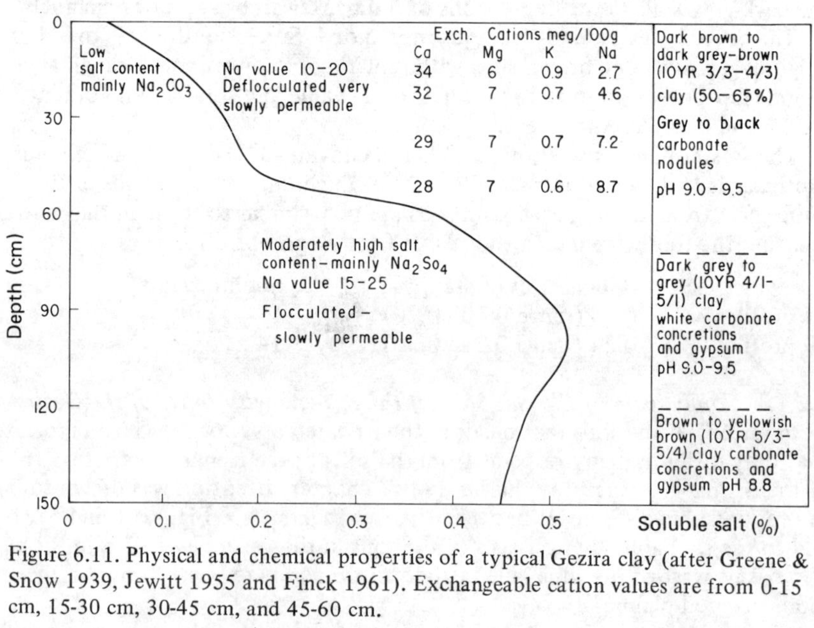

Figure 6.11. Physical and chemical properties of a typical Gezira clay (after Greene & Snow 1939, Jewitt 1955 and Finck 1961). Exchangeable cation values are from 0-15 cm, 15-30 cm, 30-45 cm, and 45-60 cm.

were formed in materials derived from various sources which were laid down under widely varying physiographic and climatic conditions. Consequently, there is a more diverse array of soils than in the Blue Nile region where materials and the depositional environment were apparently more uniform. The landforms in the White Nile trough are considered in relation to a northern zone between Kosti and Gelhak (subdivided into separate east and west bank areas) and a southern zone between Gelhak and Malakal. The soils are also discussed in relation to these distinct physical subdivisions.

1. *Northern zone east bank*

In this area there are two main assemblages of soils which occur on the recent alluvial plain rising to a level of about 382 m and on an older terrace at an elevation of about 386 m (Figure 6 and 8). The eastern border of the 386 m terrace forms the boundary between the soils formed on materials deposited by the Blue and White Niles which differ markedly in both physical and chemical properties.

The soils on the older terrace are predominantly grey clays or sandy clays which grade to grey clay at depths of 30 to 90 cm. Carbonate nodules commonly occur on the surface and in the underlying clay horizons. Particle-size analyses of composite samples from the upper 90 cm, show that clay contents range between about 40 to 60 % and that generally the coarse sand content is considerably higher than in clay soils south of Gelhak. The soils in the northern part of this zone, from Rabak to about latitude 12°20′N, are salt-affected and very strongly alkaline (pH 9.3 to 10.4). Soluble salt contents and sodium values range between 0.3-0.8 % and 26-71 respectively, and in this regard they contrast markedly with the adjacent dark grey-brown clays on Blue Nile deposits. Jewitt (unpublished report) also drew attention to the high salinity of these soils and showed that there is a concentration of salts at a depth of 30-60 cm in the profiles and a gradual decrease beneath to a depth of 180 cm. Gibb (1954) suggested that these saline-sodic soils possibly formed as a result of salt accumulation in a shallow intermittent lake under semi-arid, warm climatic conditions and a low degree of leaching.

An assemblage of soils consisting of three main groups occurs on the belt of recent alluvium which is common to both banks of the river (Figure 8) throughout the region. On the present, seasonally inundated floodplain or 'toich' are medium to heavy clay soils which are very dark grey or grey-brown in the upper 30 to 60 cm and grey or olive-grey with prominent yellow mottling beneath. Reddish iron staining along root channels is generally present in the upper horizons and carbonate nodules occur in the lower horizons of some profiles. The physical and chemical properties probably vary according to the depth and duration of flooding. Data in respect of one profile with medium acid reaction in the southern zone and averaged for the upper 90 cm are as follows:

%			Soluble salts	Na		pH	
FS	Si	C	%	value	0-30 cm	30-60 cm	60-90 cm
24	16	60	0.14	8	5.6	5.7	6.4

On old levees are grey to grey-brown sandy loams to sandy clays which are commonly underlain by sand or gravel layers at depths of 120 cm or more. The surface soil is hard and compacted when dry and graded roads along the river are generally located on these features. A few small carbonate nodules may occur in these soils on the older levees and gypsum crystals are commonly present. Soluble salt contents and sodium values are variable and reaction ranges from neutral to very strongly alkaline. The most extensive soils are grey to dark grey, medium to heavy cracking clays on old floodplains which are generally underlain by sand and gravel at depths of 120 cm or more. Chemical properties are also variable but the higher salt contents and sodium values occur in the semi-arid northern zone and there is a gradual decrease southwards.

2. *Northern zone west bank*

From Kosti to about 20 km south of Keri Kera an array of soils occurs on the Khor Abu Habl fan and related landscapes on aeolian sand deposits (Figures 4 and 8). These features occur west of and above the 382 m terrace in the belt of recent alluvium described previously. The most extensive soils are acid to mildly alkaline, reddish-brown sands and sandy loams on low dunes (qoz) and sandsheets. They have not been studied in detail but their general characteristics are described by Jewitt (1950). The qoz sands consist of 60 to 70 % coarse sand, 20 to 30 % fine sand, and 5 to 10 % clay. The yellowish red surface horizons are generally hard and compacted and grade to firm massive reddish yellow sand beneath. Soils with abrupt textural contrasts occur on sandsheets below the dunes and lower-lying gently undulating areas. Grey-brown to reddish-brown sandy loam to sandy clay loam surface horizons are underlain by brown to grey-brown clay. These soils are moderately to very strongly alkaline (pH 8.5-10.0) and they are salt-affected. Analytical data for ten profiles indicate an average sodium value of 47 (range 19-80) and a salt content of 0.4 % (range 0.1-0.8 %). On the Khor Abu Habl fan west of Keri Kera there are extensive areas of dark brown cracking clays with scattered low dunes and sandsheets. As noted by Tothill (1948), boundaries between the alluvial clays and aeolian sands are generally abrupt. The dark brown heavy clays have uniform colour and textural profiles, and analytical data for seven profiles indicate an average salt content and sodium value of 0.17 % and 16 respectively. The origin of this material has not been investigated, but it is likely that it has been derived at least in part from the weathering of volcanic rocks and not exclusively from the Basement Complex rocks of the Nuba Mountains.

6.2.3 *Southern zone*

Significant changes in landforms and soils occur south of the volcanic Jebel Ahmed Agha in latitude 11°00′N (Figure 9). On the right bank, the sandy ridge or abrupt change in slope along the eastern border of the 386 m terrace swings to the southeast around the Mabaan area and is located further from the river. Possibly as a result of this, there is a noticeable decrease in the content of coarse sand in the soils but the content of fine sand is higher. On account of the higher rainfall in this zone (600 to 800 mm) and the almost level relief of the clay plains, surface drainage is poor and there is widespread seasonal flooding and waterlogging. There are practically no salt-affected or very strongly alkaline soils and sodium values are generally less than 15. On the plains away from the belt of recent alluvium, the soils are predominantly dark grey, light to medium clays (40 to 60 % clay) but there is an extensive area of self-mulching, dark grey heavy clay soil (70 to 80 % clay) in the Melut bend. Gilgai microrelief (30 to 60 cm) is common on the seasonally flooded clay plains and a strew of grey carbonate nodules (1 to 4 cm diameter) is generally present on the crusty, weakly self-mulching surface soils. Perennial tussock grassland, mainly *Setaria incrassata* and *Hyparrhenia pseudocymbaria,* predominates in this southern zone but there are extensive areas of acacia woodland in which *A.fistula, A.seyal* and *Balanites aegyptiaca* are the dominant species.

On the west bank grey to dark grey, light to medium clay soils predominate in a 10 to 15 km belt near the river. Salt contents and sodium values are less than 0.15 % and 20 respectively. On two traverses, cross-sections (Figures 10b and c) show a gradual rise to about 390 m or more between 15 and 25 km from the river where there is a gradual change to grey-brown or dark grey-brown clay. Analytical data for six profiles indicate an average clay content of 53 %, salt content of 0.10 %, and sodium value of 10. These soils are mildly to moderately alkaline in their surface horizons.

6.3 CONCLUSION

The foregoing account of the soils of central Sudan south of the Kosti-Sennar railway-line indicates that, despite initial visual similarities in the dark cracking clays of this region, there are important physicochemical variations in the clay soils. The clays of White Nile provenance are notably more saline and alkaline than clays of presumed Blue Nile origin in the same latitude. West of the White Nile the soil pattern is complex, and reflects the greater variability in parent materials, which comprise fluviatile and lacustrine clays, aeolian sands, and colluvial sands and gravels. A recent comparison between the influence of electrolyte composition upon the structure of a Gezira cracking clay and a Texas cracking clay (Mukhtar, Swoboda & Godfrey 1974) has emphasized the practical importance of differences in soil salinity and exchangeable sodium for soil and water management practices. A thorough understanding of the age, origin, and related physicochemical properties of soils in this region is a prerequisite for future long-term agriculture in this as yet relatively undeveloped part of central Sudan.

6.4 REFERENCES

Abul-Haggag, Y. 1961, A contribution to the physiography of northern Ethiopia. Univ. London, The Athlone Press.

Abdel Salam, Y. 1966, The ground water geology of the Gezira. M.Sc. thesis, Univ. Khartoum.

Andrew, G. 1948, Geology of the Sudan. In J.D. Tothill (ed.), *Agriculture in the Sudan.* Oxford University Press.

Ball, J. 1939, Contributions to the geography of Egypt. Cairo.

Basiński, J.J. & J.H. Davies 1954, The ecology of the Jonglei area. Min. Agric., Sudan Govt.

Berry, L. 1961, The physical history and development of the White Nile. 18th Ann. Rept. Hydrobiol. Res. Unit. Univ. Khartoum. Sudan.

Berry, L. 1962, Large scale alluvial islands in the White Nile. *Rev. Géomorph. dyn.* 12: 105-8.

Blandford, W.T. 1869, On the geology of a portion of Abyssinia. *Quart. J. Geol. Soc. Lond.* 25:401-406.

Bunting, A.H. & J.D. Lea 1962, The soils and vegetation of the Fung, east central Sudan. *J. Ecol.* 50:529-58.

Butzer, K.W. & C.L. Hansen 1968, Desert and river in Nubia. Univ. of Wisconsin Press, Madison.

Butzer, K.W., G.L. Isaac, J.L. Richardson & Celia Washbourn-Kamau 1972, Radiocarbon dating of East African lake levels. *Science* 175:1069-76.

El-Attar, H.A. & M.L. Jackson 1973, Montmorillonitic soils developed in Nile River sediments. *Soil Sci.* 116:191-201.

Finck, A. 1961, Classification of Gezira clay soil. *Soil Sci.* 92:263-7.

Flint, R.F. 1959, Pleistocene climates in east and southern Africa. *Bull. Geol. Soc. Am.* 70:343-74.

Gibb, Sir Alexander & Partners 1954, Estimation of irrigable areas in the Sudan 1951-3. Rep. to Sudan Govt. London, Metcalfe & Cooper Ltd.

Grabham, G.W. & R.R. Black 1925, Report of the mission to Lake Tana, 1920-21. Cairo.

Greene, H. & O.W. Snow 1939, Soil improvement in the Sudan Gezira. *J. Agric. Sci.* 29: 1-34.

Gunn, R.H. 1955, The use of aerial photography in soil survey and mapping in the Sudan. *Soils and Fert.* 18:104-6.

Gunn, R.H. 1974, A soil catena on weathered basalt in Queensland. *Aust. J. Soil Res.* 12: 1-14.

Gunn, R.H. & R.W. Galloway (1978), Silcretes in south-central Queensland. Monograph Series, Univ. of New England, Armidale.

Hurst, H.E. 1952, The Nile. Constable, London.

Jewitt, J.N. 1950, Goz soils of the Anglo-Egyptian Sudan. Trans. 4th Int. Cong., *Soil Sci.* 1:285-8.

Jewitt, T.N. 1955, Gezira soil. Min. Agric. Sudan Govt., Bull. 12.

Jewitt, T.N. & K.R. Middleton 1955, Changes in Sudan Gezira soil under irrigation. *J. Agric. Sci.* 45:277-9.

Jones, T.A. 1948, Preliminary report on soil and vegetation survey undertaken in connection with the Oxford University Soil Survey expedition. Unpubl. Sudan Govt. Rept.

Kendall, R.L. 1969, An ecological history of the Lake Victoria basin. *Ecol. Monog.* 39: 121-176.

Mohr, P.A. 1971, The geology of Ethiopia. Haile Sellassie I University Press, Addis Ababa (reprint, with expanded foreword, of 1962 edition).

Morin, W.J. & W.T. Parry 1971, Geotechnical properties of Ethiopian volcanic soils. *Geotechnique* 21:223-32.

Mukhtar, O.M.A., A.R. Swoboda & C.L. Godfrey 1974, The effect of sodium and calcium chlorides on structure stability of two vertisols: Gezira clay from Sudan, Africa, and Houston Black Clay from Texas, USA. *Soil Sci.* 118:109-119.

Nilsson, E. 1953, Contribution to the history of the Blue Nile. *Bull. Soc. Géogr. d'Egypte* 25:29-47.

Oakes, H. & J. Thorp 1950, Dark clay soils of warm regions variously called Rendzina, Black Cotton Soils, Regur, and Tirs. *Soil Sci. Soc. Amer. Proc.* 15:347-54.

Shukri, N.M. 1949, The mineralogy of some Nile sediments. *Quart. J. Geol. Soc.* 105: 511-34.

Sudan Government 1929, Gezira cross levelling (CAD 245).

Sudan Government 1950-51, Kenana soil survey.

Tothill, J.D. 1948, A note on the origins of the soils of the Sudan. In J.D. Tothill (ed.), *Agriculture in the Sudan.* Oxford University Press.

United States Department of Agriculture 1960, Soil classification, a comprehensive system. 7th Approximation. Washington.

Whiteman, A.J. 1971, Geology of the Sudan Republic. London, Oxford University Press.

Williams, M.A.J. 1966, Age of alluvial clays in the western Gezira, Republic of the Sudan. *Nature* 211:270-1.

Williams, M.A.J. & D.A. Adamson 1973, The physiography of the central Sudan. *Geog. J.* 139:498-508.

Williams, M.A.J. & D.A. Adamson 1974, Late Pleistocene desiccation along the White Nile. *Nature* 248:584-6.

MARTIN A.J.WILLIAMS / DONALD A.ADAMSON
Macquarie University, North Ryde, Australia
HASSAN HAG ABDULLA
Gezira Research Station, Wad Medani, Sudan

7

Landforms and soils of the Gezira: A Quaternary legacy of the Blue and White Nile rivers

7.1 BACKGROUND TO THE PRESENT STUDY

The Gezira is a triangular wedge of land bounded by the Blue Nile to the east, the White Nile to the west, and by the rising ground of the Manaqil Ridge to the south (Figures 1 and 2). A convenient albeit arbitrary southern boundary is the Sennar-Kosti railway line, which runs through a spectacular landscape of granitic and metamorphic inselbergs or jebels bounded by sweeping vistas across extensive grassy plains. 'Gezira' is Arabic for 'island', and the Sudan Gezira plain is more than a metaphorical island between two of the world's mightiest rivers, for it owes its very existence to the depositional activities of those rivers in the past (Tothill 1946, 1948).

Any investigation into the age and origins of the Gezira begins with the realisation that the Gezira is the economic powerhouse of the Sudan (Gaitskell 1959). Although small relative to the country as a whole (the area occupied by the Gezira is barely 1 % of the total area of 2.5 million km^2), the Gezira provides about 65 % of the export revenue of the Sudan (National Council for Research 1975). Half of Sudan's exports stem from cotton, and over 75 % of this cotton is grown on irrigated clay soils in the Gezira. About 30 % of the world's long-staple cotton comes from the Sudan, so that attempts to understand the factors that influence crop yield in the Gezira are of major concern within and beyond the Sudan.

Although often described as a flat and featureless clay plain, the Gezira is neither totally flat nor entirely made up of clay. Sinuous depressions flanked by low sandy ridges radiate out north and northwest across the Gezira plain from about the latitude of Sennar. Recognised in 1963-64 as former courses of the Blue Nile by Dr Robert Smith, soils consultant, and by M.A.J. Williams, reconnaissance soil surveyor with Hunting Technical Services, certain of these palaeochannels and their associated soils were mapped in detail by Williams during 1962-64.

It was during two seasons of mapping soils, vegetation and land-use between the Blue and White Nile, and more especially along the right bank of the White Nile in 1963-64, that certain problems worthy of more detailed study began to emerge (Hunting Technical Services 1964b; Williams 1966).

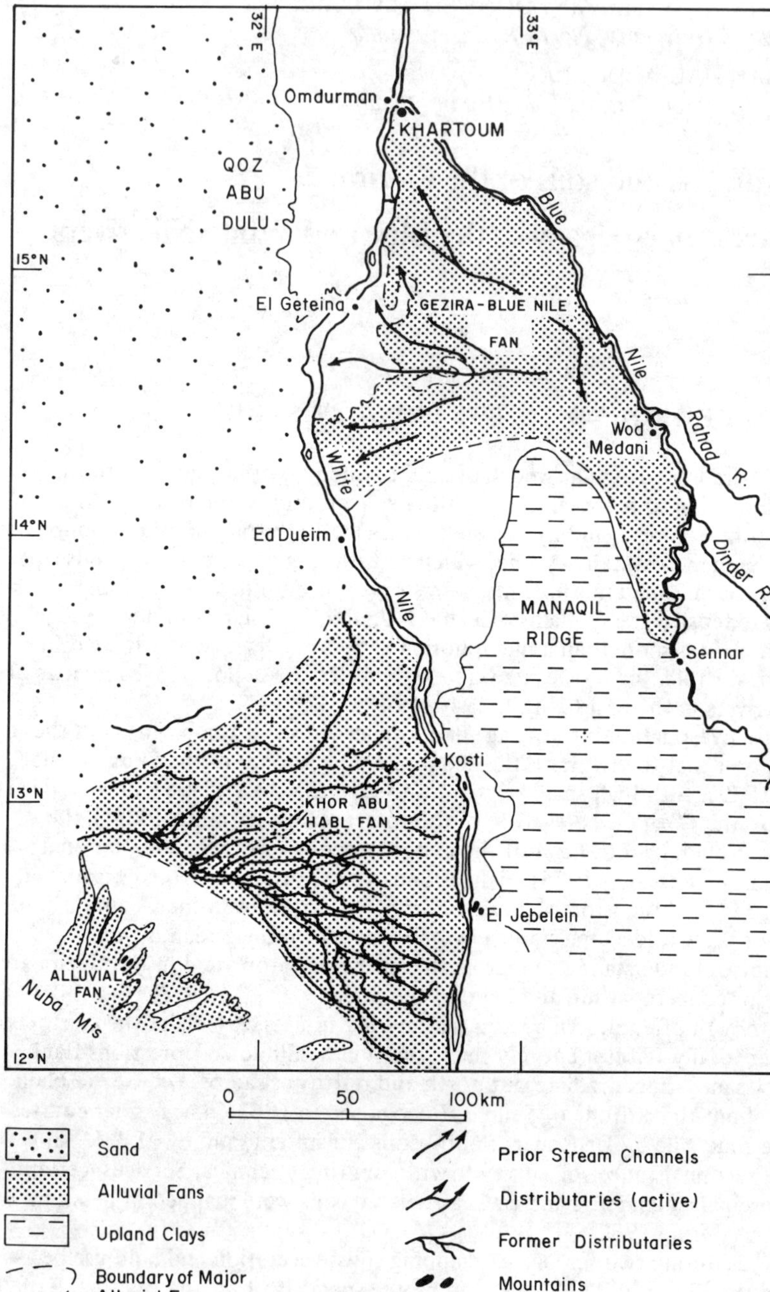

Figure 7.1. Major Physical Features, Central Sudan.

Neolithic and older prehistoric sites were discovered on many of the sand ridges in the western Gezira. Abundant pottery fragments, stone rings and polished and flaked stone implements were found at Jebel Tomat between the two low hills that comprise the 'twin' jebels. Concentrations of unbroken freshwater and amphibious shells at depths of one to two metres beneath the present surface indicated that the White Nile had on occasion stood at a level several metres above its modern flood level between about 12 000 and 8 000 years BP.

Of more immediate practical concern was the realisation that the pattern of near-surface salinity immediately west of the White Nile had little to do with the present climate and a great deal to do with late Quaternary behaviour of the White Nile (Williams 1968b). Similarly, the 100 000 hectares of sandy ridges and clay-floored swales in the northwestern Gezira reflected changes in the flow regime and former courses of the Blue Nile as well as fluctuations in the level of the White Nile (Williams 1966, 1968a).

Except for the now classic excavations by Wellcome and colleagues at Jebel Moya over 60 years ago (Addison 1949; Mukherjee, Rao & Trevor 1955), the brief study by Balfour Paul (1952) at Qoz Kabaro and elsewhere in the Gezira, and sparse surface collections along the White Nile, the country south of Khartoum was archaeologically unknown. It was clear that changes in the flow regime of the Blue and White Nile rivers must have had a profound influence upon the prehistoric hunter, fisher and gatherer communities and early cultivators and pastoralists in this part of Africa. And so the present study began.

After two reconnaissance trips by Williams & Adamson in January and December 1972, during which we began joint work with colleagues from Khartoum and Wad Medani (Williams & Adamson 1973, 1974; Williams et al. 1974), an archaeological survey party of three Berkeley graduate students directed by Professor J.D. Clark began joint fieldwork with us at Jebel Tomat and at Shabona prehistoric site early the following year. The aim of this chapter is to describe the present Gezira landscape. In so doing, it will become apparent that almost all the major landscape features are a legacy from late Quaternary times, and that during prehistoric settlement of the Gezira the landscape was continuously evolving and was in a far greater state of geomorphic flux than is the case today.

7.2 GENERAL DESCRIPTION OF THE GEZIRA

We have already noted in chapters 4 and 5 that the irregular topography of the Basement Complex landscape was partly buried beneath sands and gravels of Cretaceous age (Nubian Sandstone), to be submerged once more by the alluvial fan deposits carried down from Ethiopia during the past 20 to 30 million years of the later Tertiary and Quaternary. The Gezira itself is in part a vast clay-veneered low-angle alluvial fan, but its clay-mantled uplands to the south are in part erosional. Indeed, in our view, which differs from that put forward by Gunn in chapter 6, it is probable that many of the clay plains

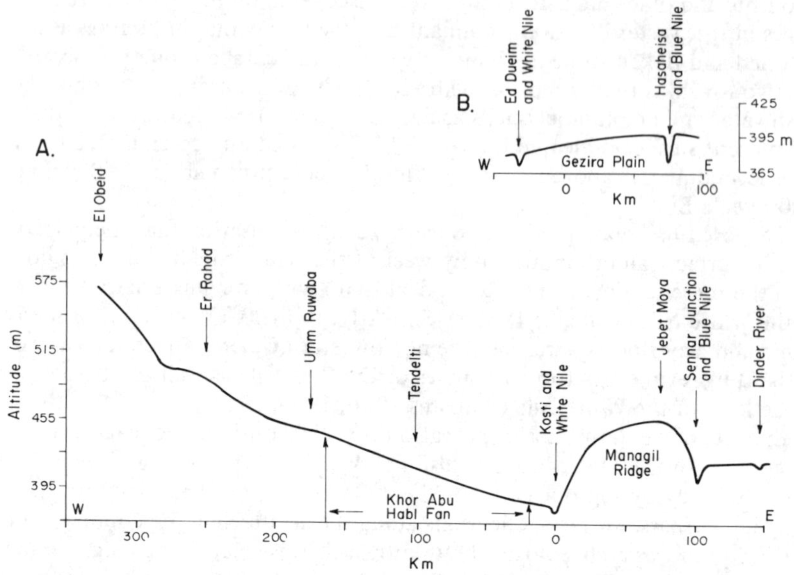

Figure 7.2. East-west section along the Sennar-El Obeid railway-line across the Gezira and Khor Abu Habl alluvial fans.

of western and southern Sudan have developed by prolonged *in situ* weathering of the Tertiary basalts and of the ancient Basement Complex metasediments and associated intrusive and extrusive rocks that form the bedrock foundation of much of the Sudan. The distinction between residual upland clays and depositional lowland clays is an important one agriculturally as well as geomorphically, for the residual clays tend to be far more permeable and far less saline than their alluvial counterparts (Bunting & Lea 1962; Williams 1968c).

The major physical features of central Sudan show up very clearly on ERTS satellite imagery, of which Figure 1 is a simplified interpretation. Both the Gezira fan and the Khor Abu Habl fan west of the White Nile (Figures 1 and 2) emerge from and are flanked by uplands with variable proportions of rocky outcrops and clay-mantled slopes. At a slightly lower level than the fans, but sometimes merging with them, is the alluvial flood-plain of the White Nile, only part of which is active today. Banked up against the Abu Habl fan to the north and interdigitating with the older alluvial clays of the White Nile to the east are the extensive sand sheets and fixed linear dunes of eastern Kordofan. Closer inspection of the Gezira fan (Figure 3) shows that it slopes gently northwest from 405 m at Manaqil to below 380 m near Khartoum, a fall of 25 m in 170 km or 6 $cm.km^{-1}$. Before giving more details about the major landforms in the Gezira, it will be useful to discuss the two rivers that created these landforms.

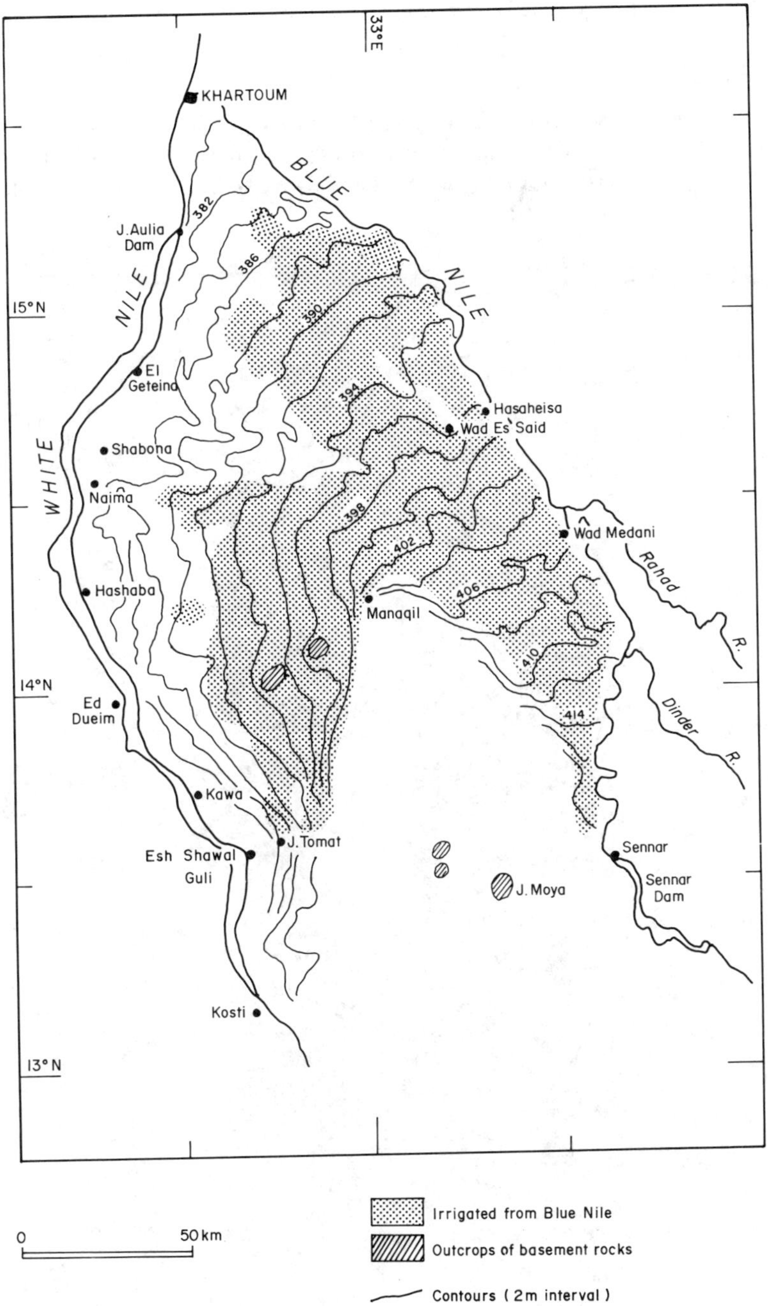

Figure 7.3. Contour map of the Gezira.

7.3 THE BLUE AND WHITE NILE RIVERS

7.3.1 *The Blue Nile*

The Ethiopian headwaters of the Blue Nile (or Little Abbai) rise at an elevation somewhat above 1 850 m and flow north for about 100 km to join Lake Tana at 1 829 m above sea-level (Figures 4 and 5). Lake Tana is a broad shal-

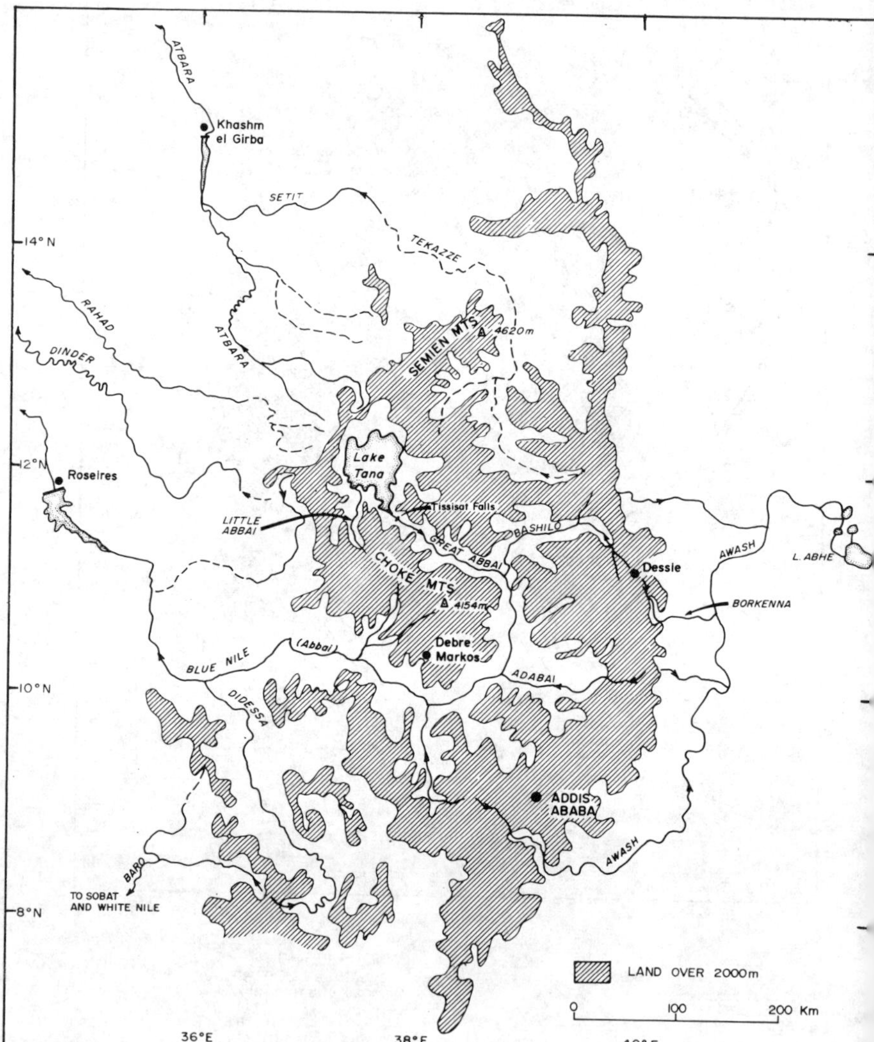

Figure 7.4. The Blue Nile headwaters.

low lake up to 14 m deep, up to 67 km wide and up to 78 km long (Rzóska 1976). The age of the lava dam at the southeast corner of the lake where the Blue Nile (or Great Abbai) flows out from Lake Tana is not yet known, but may be Pliocene or younger. From Lake Tana the river flows roughly southeast for 30 km until the Tissisat Falls, where it drops 50 m. Below the Falls the river enters a canyon which deepens progressively downstream, attaining a depth of roughly 1 500 m and a width up to 30 km downstream of the Debre Markos bridge.

During its descent from the Falls, the river flows in turn southeast, south, southwest and finally north of west around the volcanic massif of the Choke Mountains, the highest point of which is at 4 413 m (Figure 4). Such a course is curious, particularly when it is seen that the west-flowing Bashilo river, a major left bank tributary of the Abbai, originates on the rim of the eastern scarp south of Dessie, only a few kilometres from the southeast-flowing Borkenna river, an important tributary of the Awash. It is possible that much of the original drainage which once flowed east from the Semien and Choke highlands, including the Tekazze headwaters, became captured and diverted westwards as a result of the earth movements related to the formation of the Afar rift. Faulting of the Trap Series lavas, reactivation of older zones of weakness in the underlying Basement Complex rocks, and selective erosion along SE-NW and SW-NE trending faults, fissures and major joints by the oft-rejuvenated Abbai and its companion river the Tekazze created the drainage pattern we now see in the eastern Sudan.

Between Lake Tana and the Sudan border the Blue Nile descends 1 350 m in about 850 km, a fall of roughly 1.6 m every kilometre (Figure 5). Between the border and Roseires the river has cut itself a bedrock channel up to 40 m deep. At Roseires the river emerges from its straitened valley and begins to meander across the proximal sector of its alluvial fan. When full the reservoir at Roseires is at 480 m above sea-level, or about 60 m above the level of the Sennar reservoir some 280 km further north. Between Sennar and Khartoum the Blue Nile falls a further 65 m over a distance of about 360 km (Figure 6). The total fall in river level between the Sudan border and Khartoum is about 126 m (from 500 to 374 m) over a distance of about 900 km, or roughly 13 cm every kilometre, which is 12 times less steep than the mean channel gradient between Lake Tana and the Sudan border over an equivalent distance of flow (Figure 5).

The practical significance of these contrasts in gradient is obvious. A river's ability to carry its load, here defined as the size of the largest particle it is capable of transporting as bedload, increases exponentially with river velocity, which itself increases with channel gradient. As an approximate rule of thumb, doubling the river velocity increases the size of the largest fragment carried by a factor of 50. Conversely, with any slackening of velocity very coarse material will be deposited forthwith. Accordingly, much of the coarser bed-load material eroded from the granites, sandstones and basalts of Ethiopia now probably forms the base of the Gezira Formation. The suspended load, made up at present of roughly equal parts of clay, silt and sand-sized particles, is mostly swept on through Sudan to Egypt.

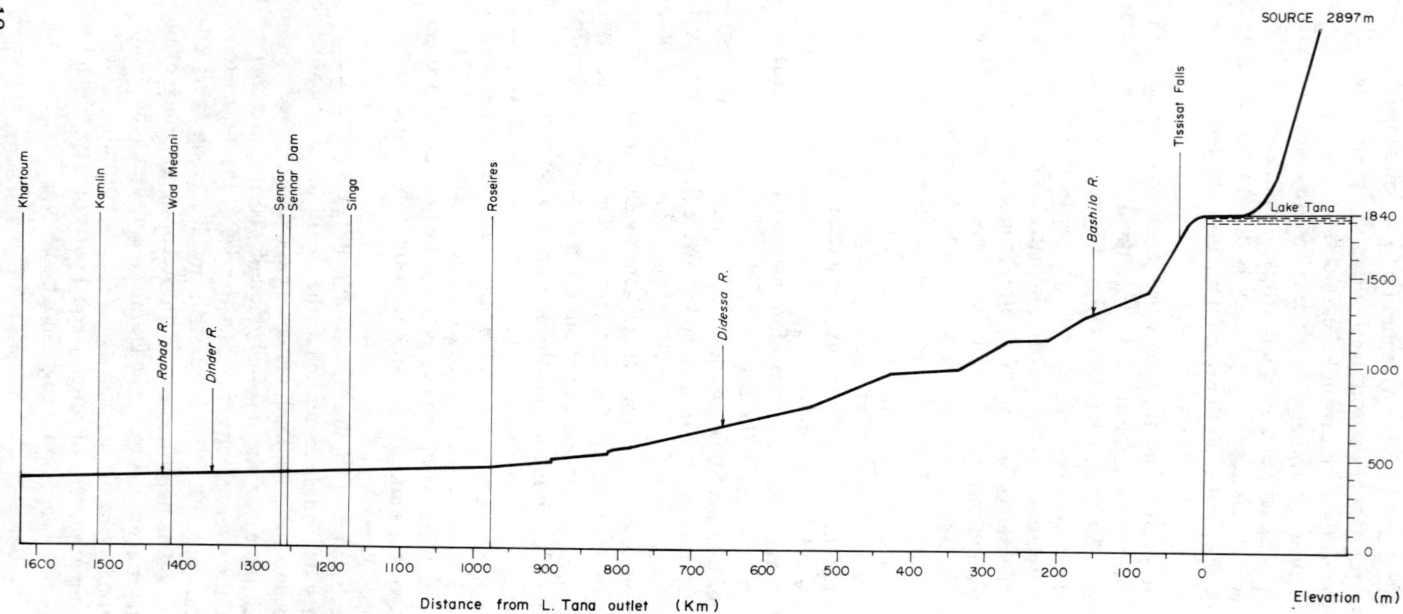

Figure 7.5. Longitudinal profile of the Blue Nile in Ethiopia.

The Blue Nile is highly seasonal, with a ratio of peak flow to low flow of 40:1, and a peak sediment concentration at Khartoum of 4 000 mg.l^{-1} in August as against only 100 mg.l^{-1} in June (Badri 1972 and Table 1). Such seasonality is directly related to the rainfall regime in the Ethiopian uplands, where 90 % of the annual precipitation of up to 2 000 mm of rain falls between June and October, with a peak during July and August.

Mean annual discharge over the past 50 years is 50 km^3 for the Blue Nile at Roseires, of which overflow from Lake Tana contributes only 4 km^3 or roughly 7 % (Rzóska 1976). The seasonal Rahad and Dinder rivers which join the Blue Nile near Medani contribute a further 1 and 3 km^3 respectively. Seepage and irrigation losses reduce Blue Nile flow to 50 km^3 at Khartoum. The total amount of sediment contributed on average each year by the Blue Nile to the main Nile at Khartoum was 40 million tons before completion of the Roseires Dam in 1965. With a peak flow to low flow ratio of 40:1, and an August/June sediment concentration of 4 000 and 100 mg.l^{-1} respectively, it is easy to see why the Blue Nile carries up to 1 600 times more sediment during the summer floods than at the end of the dry season (Table 1). The long-term consequences for Egypt, eloquently documented by Toynbee, have been common knowledge since well before Herodotus visited Lower Egypt over 2 400 years ago.

Nor should the role of the Tekazze or Atbara be forgotten, for with a mean annual flow of 12.5 km^3 it furnishes an additional 14 million tons of sediment during the flood season, drying out completely between floods. The Blue Nile and Atbara between them provide 90 % of the total flood discharge of the main Nile, of which 68 % is from the Blue Nile and 22 % from the Atbara (Table 1). These same two rivers provide 96.5 % of the Nile's sediment load, with 72 % from the Blue Nile, and 24 % from the Atbara. Interestingly enough, the Blue Nile contributes only 17 % of the Nile's low water discharge, and the Atbara nothing. The outstanding 83 % comes from the White Nile, and we will shortly explore some of the consequences of this very important hydrological fact (see chapter 9).

Table 7.1. Present hydrological regime of the Nile.

River	Discharge km^3	Sediment load (million tons)	% of peak flow	% of low flow
White Nile	27.5	2	10	83
Blue Nile	51	41	68	17
Atbara River	12.5	14	22	0
Main Nile	91	57		

a) Ratio of peak flow to low flow:
White Nile 5:2,
Blue Nile 40:1,
Main Nile 16:1.
b) Note: Blue Nile sediment concentration:
June: 100 mg/l,
August: 4 000 mg/l.

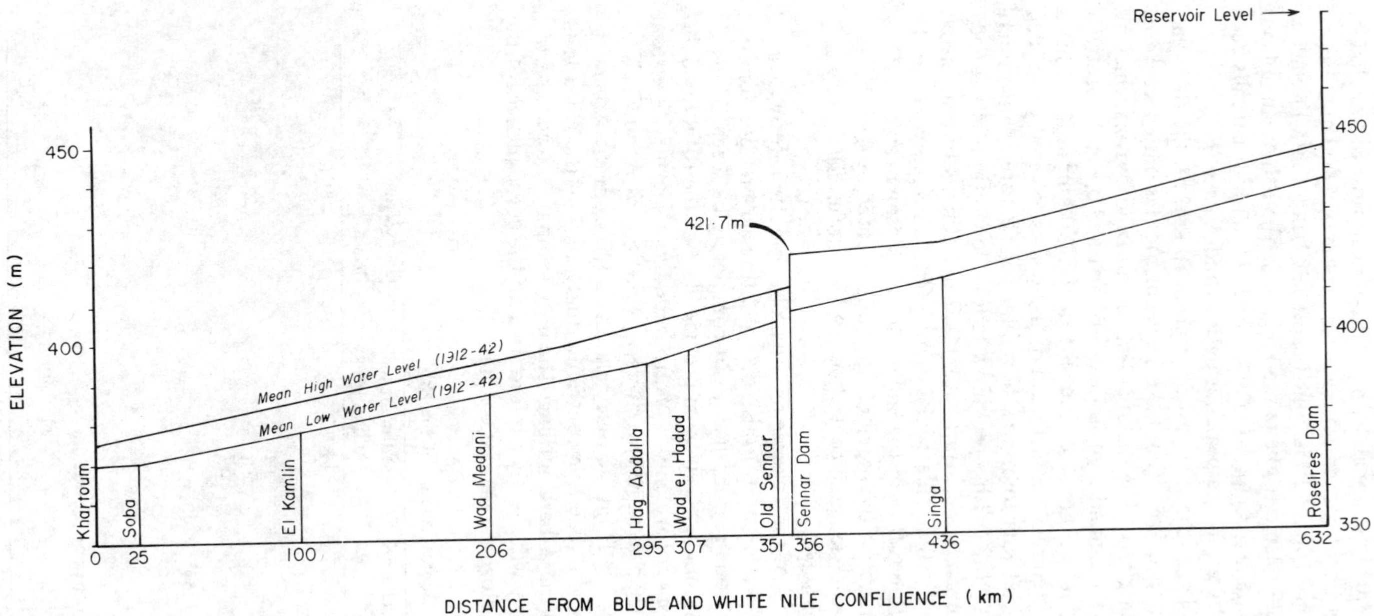

Figure 7.6. Longitudinal profile of the Blue Nile in Sudan (after Hunting Technical Services 1966).

7.3.2 *The White Nile*

It is the addition of the Ugandan rivers to the main Nile via the White Nile that make it the longest river in Africa, totalling 6 670 km from the headwaters of the Kagera river to the Mediterranean Sea. The combined catchment areas of the Blue Nile, the White Nile and the main Nile make the Nile basin the fourth largest river basin in the world, after the Amazon, the Zaire, and the Mississippi-Missouri river basins.

The upper Kagera originates from swamps in Burundi at slightly over 2 200 m above sea-level, falling swiftly to Lake Victoria at 1 100 m elevation. From Lake Victoria to Khartoum the White Nile falls 720 m over a distance of roughly 3 000 km, in strong contrast to the Blue Nile which descends twice as far in half the distance. Regulated by Lakes Victoria and Albert and by the extensive swamps of the southern Sudan the White Nile is far less seasonal than the Blue, with a ratio of mean peak flow to mean low water flow of 5:2. In the words of Sir Samuel Baker (1866), the Blue Nile 'is a rapid mountain stream, rising and falling with great rapidity', and the White Nile is a river of lake origin, flowing through vast treeless swamps in a region of 'malaria, marshes, mosquitoes, misery'. In fact, the upland swamps of Uganda are a hidden blessing, for they damp down the flood peaks and sustain the perennial flow of the White Nile (Thompson 1976), which provides the main Nile with 83 % of its low season flow.

From Lake Victoria the Nile flows north through the submerged river valley occupied by Lake Kyoga, over the Murchison Falls, and on into Lake Albert, only to emerge a few kilometres from its point of entry as a sluggish river flowing in the northern end of the Albert Rift until it reaches the Aswa Wrench Fault, where it takes a right-angled turn to the north through the dissected uplands straddling the border with the Sudan. Aptly named the Bahr el Jebel (or river of the mountain) in its upper reaches in southern Sudan, the White Nile flows across its alluvial fan by the time it reaches Juba, upstream of the swamps of the Sudd. The Bahr el Jebel, fed from Lake Albert, Lake Victoria, and from strongly seasonal rivers in southern Sudan, has an annual discharge slightly over 23 km^3 before it enters the Sudd, with a subdued flood peak relative to the rest of the Nile system. Its discharge at Lake No, where the Bahr el Ghazal enters from the west, is barely 10 km^3 (Hurst & Phillips 1931). Over half the original volume of water in the Bahr el Jebel is therefore lost by seepage and evapo-transpiration in and around the swamps, yet oddly enough the total salt concentration in water leaving the Sudd is almost the same as that entering it (Talling 1957, 1976; Bishai 1962). The Sudd is functioning as a gigantic chemical filter, as well as sediment trap (Williams & Adamson 1973, 1976).

The Bahr el Ghazal joins the White Nile north of the main Sudd bringing a further 3 km^3 of water, to which the Sobat adds another 10 km^3 at its junction above Malakal, bringing the total White Nile flow to 23 km^3 The annual sediment load of only 2 million tons contributed to the main Nile at Khartoum indicates that volume for volume the Blue Nile is able to carry over ten times more sediment than the White Nile, and transport over 20 times more in fact. Doubtless much of the coarser material that could be

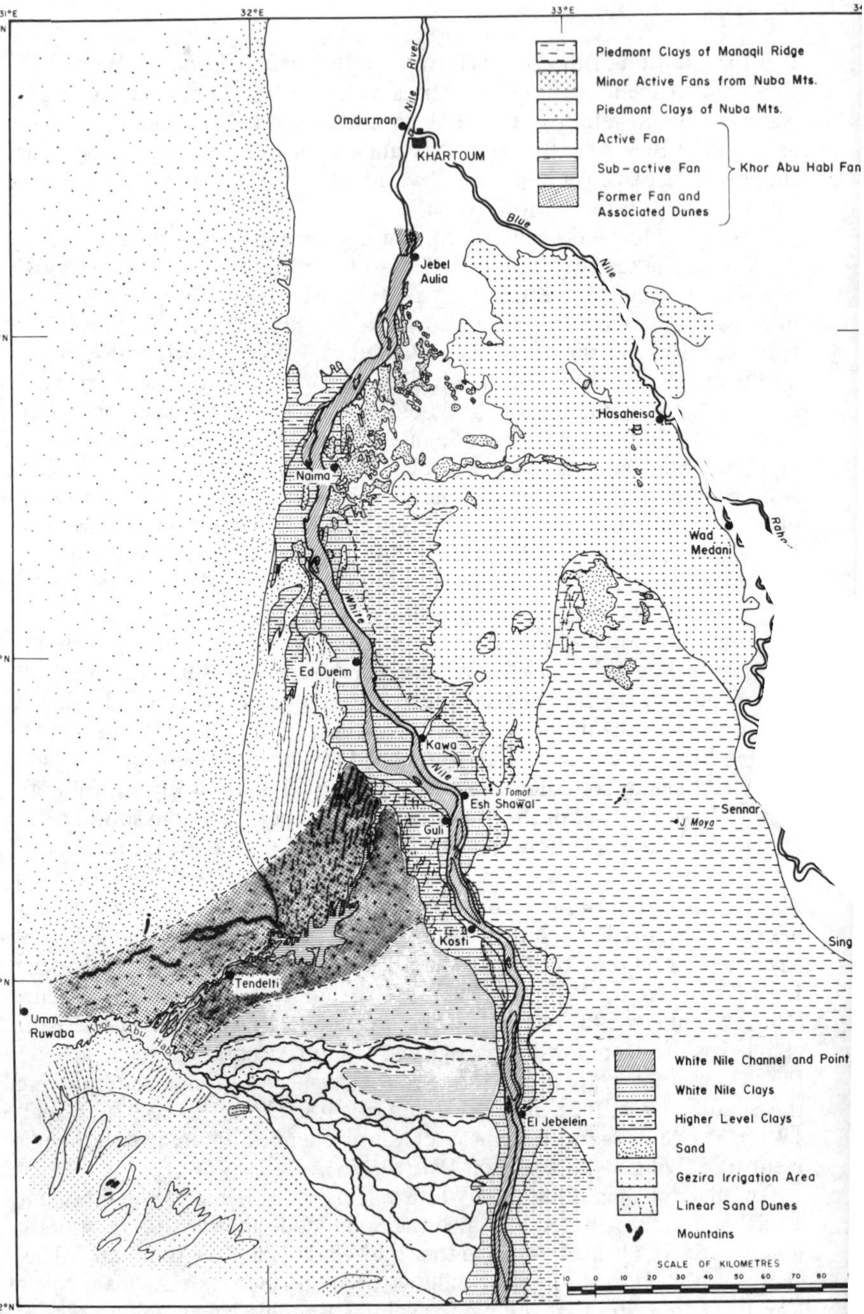

Figure 7.7. The Blue Nile and Khor Abu Habl Alluvial Fans, Central Sudan.

transported by the White Nile is deposited in the Ugandan lakes and swamps and the finer suspended material is filtered out in the Sudd or decanted out onto the seasonally-flooded 'toichlands' between Malakal and Renk.

The White Nile falls from 386 to 374 m in the 810 km between Malakal and Khartoum, equivalent to about 1 $cm.km^{-1}$. The Blue Nile by contrast falls 12 $cm.km^{-1}$ between Roseires and Singa, and 9-10 $cm.km^{-1}$ between Sennar and Khartoum at high water (Hurst & Phillips 1931). A further distinction between the two rivers is their channel form. The Blue Nile is a deep, narrow, sinuous river entrenched into its former flood plain. The White Nile is broad and shallow with a tendency to bifurcate around the elongated mid-channel bars or islands common along its slightly incised lower reaches. Unlike the Blue Nile, the present White Nile below Malakal is not particularly sinuous, and is better described as straight than as either braided or meandering. We will now turn to the landforms created by these two rivers upstream of their junction at Khartoum.

7.4 LANDFORMS OF THE GEZIRA

7.4.1 *Age and origin of the Gezira plain*

The Gezira is a gently sloping clay plain dotted with occasional prominent rocky hills in the south and interspersed with low sandy rises in the north. Until about 30 years ago many geologists and natural scientists thought that the clays had originated as wind-blown dusts, although whether of northerly or southerly provenance was not too clear.

It was not until 1946 that Tothill proved that the clays were in large measure alluvial and not aeolian. The distribution of freshwater, amphibious and terrestrial mollusca in the top 180 cm of the Gezira clays showed that these clays had been laid down by formerly more extensive floods from the Blue and White Nile rivers (Tothill 1946, 1948). Tothill's estimate of 10 000 years for the age of these alluvial clays was a remarkably accurate prediction, verified by our radiocarbon dating of the sub-fossil mollusca (Williams & Adamson 1973, 1976 and chapter 9). Principal of the College that became the University of Khartoum, Director of Agriculture, distinguished scientist and polymath, J.D. Tothill's greatest legacy to science and culture in the Sudan is the still unsurpassed volume that he inspired and edited on 'Agriculture in the Sudan'. In this volume Tothill (1948) drew attention to a high strandline of the White Nile etched into a sand-ridge at Hashaba at an elevation of 381.2 m. Dated shell-beds collected from soil-pits near Tagra and from slightly east of Esh Shawal (Williams 1966) show that the White Nile was at least 3 m higher than its present unregulated flood level towards 11 500-11 000 BP and towards 8 500-8 000 BP, again vindicating the accuracy of Tothill's observations.

7.4.2 *The clay uplands*

Not all the clays in the Gezira are alluvial, for above about 386-400 m the alluvial clays merge into the colluvial and sedentary clays developed over the Basement Complex and younger rocks of the Managil Ridge (Bunting & Lea 1962). The approximate lower limit of these autochthonous piedmont clays is shown in Figure 7, which is based on detailed interpretation of the ERTS imagery checked by extensive ground control between the two Niles, and by more restricted field observations west of the White Nile.

Sedentary clays differ from alluvial clays in three main respects. They lack fluviatile sedimentary structures, contain no sub-fossil freshwater or amphibious mollusca, and grade progressively into weathered bedrock. Around the periphery of the clay uplands is a broad transitional belt not shown on our figure in which alluvial clays may overlap the weathered Basement rocks and their derivatives. A good example of a polygenic section was seen by Williams during the digging of a well at Ras Ras village some 15 km south of Jebel Tomat early in 1964. The section comprised 2 m of grey-brown sandy clay overlying 2.5 m of light brown clayey sand with abundant large concretions (up to 12 cm across) of carbonate-cemeted very fine sand. Below a depth of about 4.5 m the calcareous sand graded into a multi-hued and highly weathered bedrock, with gneissose banding clearly evident from 18 m onwards (Hunting Technical Services 1964).

Andrew (1944) has suggested that the 'final stage of infilling of depressions' (in the Blue and White Nile valleys) 'is an alluvial sandy layer patchily cemented by freshwater limestone immediately underlying the clay'. At over 300 two or three metre boreholes and pits with a surface horizon of clay or sandy clay, half of the basal horizons (usually below 140-160 cm) were sands or clayey sands (Hunting Technical Services 1964), so that at least along the eastern side of the White Nile true alluvial clays are quite thin, in contrast to the often very deep sedentary and colluvial clays of the Managil Ridge.

7.4.3 *The clay lowlands*

In an area in which the ruling gradients are a few centimetres per kilometre, terms like 'upland' and 'lowland' are of course very relative. The rocky jebel that is a key landmark to a nomad in east-central Sudan would scarcely be noticed among the granitic highlands of western Eritrea. Nevertheless, the distinction between clay upland and clay lowland is important in practice, and is as well understood by the peasant farmer as by the plant ecologist, soil surveyor, or irrigation engineer.

Figure 7 also distinguishes between 'White Nile clays' and 'higher-level clays'. Although these two alluvial formations were undoubtedly laid down by former White Nile floodwaters, the higher-level clays are somewhat older, and may well have received some material from certain of the former Blue Nile distributary channels shown in Figure 1. As a first approximation, we may consider the boundary between White Nile and higher-level clays as delineating the maximum extent of flooding by the White Nile at the start of the Holocene Period some 11 000 years ago (Williams & Adamson 1976 and chapter 9). The White Nile clays are deeper, darker, and heavier-textured than their higher-level counterparts, but both formations merge northwards into a complex of fixed sand dunes and clay-floored swales.

7.4.4 *Palaeochannels and source-bordering dunes*

Satellite imagery and air photo mosaics show that the Khor Abu Habl alluvial fan is criss-crossed with the traces of former distributary channels and their source-bordering dunes. Similar but less numerous former channels are also evident from the air throughout the length of the Gezira fan, though often obscured on the ground by the effects of half a century of intensive canal-construction and irrigation (Figure 1). At certain localities in the Gezira such as the qoz near Wad es Said (14°40′N, 33°13′E) current-bedded coarse sands with unbroken concentrations of the amphibious mollusc *Pila wernei* testify to former water flow along these now defunct channels. The

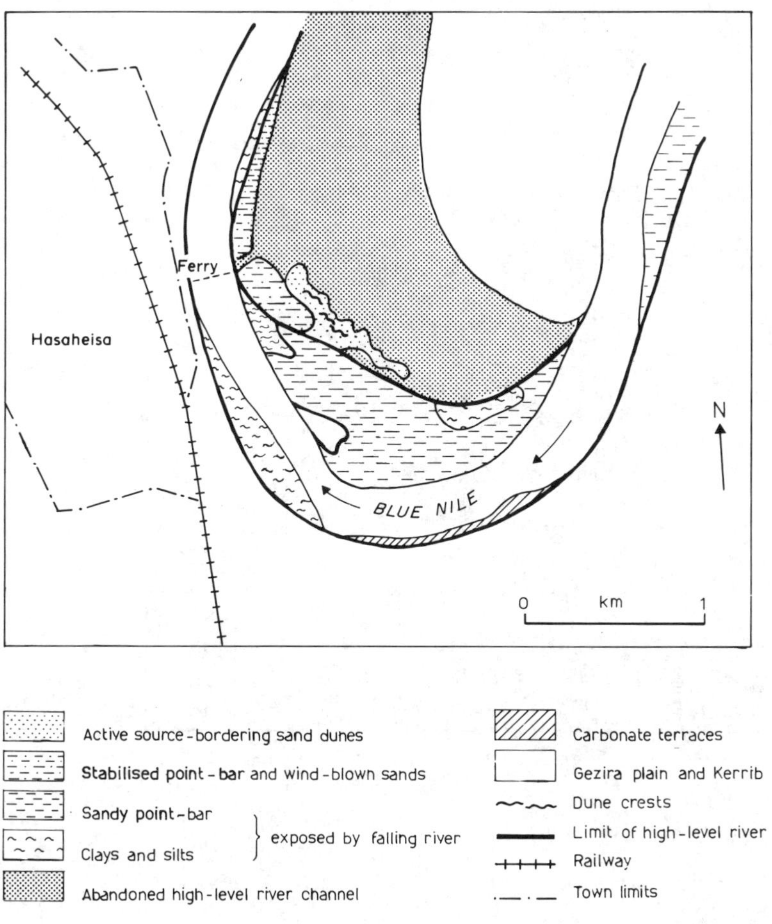

Figure 7.8. The active source-bordering dunes alongside the Blue Nile opposite Hasa Heisa.

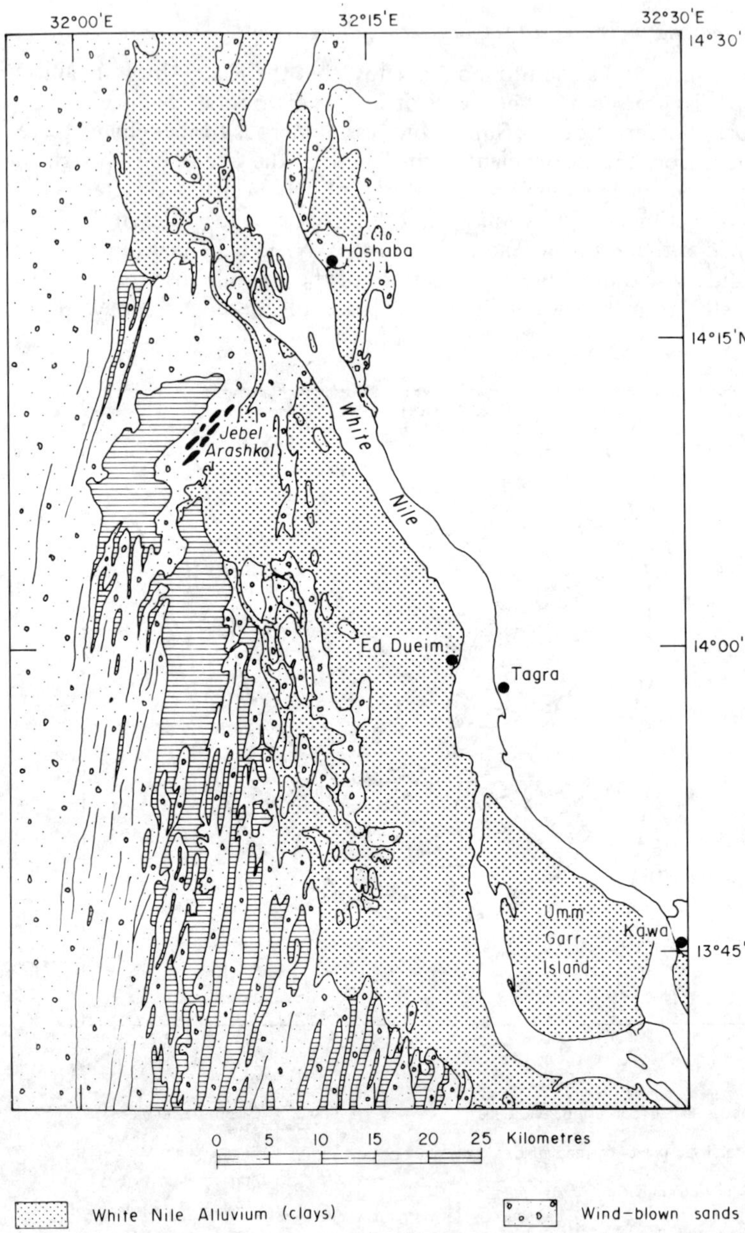

Figure 7.9. The Sand Dune Complex west of the White Nile.

mineral assemblage of these alluvial sands includes augites, pyroxenes, agates and zeolites of probable Ethiopian provenance, and all of the channels converge upstream upon the present Blue Nile (see chapter 9, Figure 3).

Immediately to leeward of many of these prior channels are prominent aeolian sand dunes. The dunes are located on the eastern, downwind margin of the channels, and are typical lee source-bordering dunes. They lack the alluvial structures and mollusc fauna of the original channel sands, but are intimately associated with the channels. Important tombs (qubbas) and village mosques are often built upon these now stable dunes, and their surface concentrations of prehistoric stone implements, pottery and broken shells testify to the attraction they have long held for man in this region.

Prerequisites for the formation of source-bordering dunes are a constantly replenished supply of alluvial sand, seasonal exposure of the channel floor, seasonal strong winds, and a sparse or absent shrub and tree cover along the channel banks. At the present time conspicuous lee dunes are developing from seasonally-exposed point-bar sands in the Blue Nile channel opposite Hasaheisa (Figure 8). Judging from the position of the ancient lee dunes relative to their ancestral parent channels, it seems that the dominant sand-moving winds were from the southwest. Since the dry season winds now blow from the north and northeast, and southerly winds today herald the Blue Nile floods and the wet season, the lee dunes were probably built up immediately before the Blue Nile floods, when water level was lowest, before northward movement of the intertropical convergence zone brought summer rain to the central Sudan.

7.4.5 *Linear dunes and sand-sheets*

The source-bordering dunes that parallel the now defunct distributary channels that radiate northwest across the Gezira from about the latitude of Wad Medani (Figure 1) are a special category. More extensive are the linear dunes and sand ridges that border the White Nile right bank from Jebel Aulia southwards to Hashaba (see Figures 7 and 9). The meridional trend of these linear dunes is clearly defined on the left bank of the White Nile, particularly to the south and west of Ed Dueim (Figure 9). The swales between these linear dunes are lined with alluvial clay. In chapter 9 we will see that the dunes were active during the arid terminal Pleistocene and became fixed during the wetter early Holocene. In fact, the dunes are partly drowned in a sea of White Nile alluvial clay. During the dramatic and probably rapid rise in the flood level of the White Nile towards 11 500-11 000 years ago, the riparian flanks of the dunes became eroded by wave action, and arcuate embayments overlooked by locally oversteepened dunes came into being. It is likely that littoral foredunes developed along both margins of the expanded White Nile at about this time. The linear dunes, foredunes and wave-modified littoral sand ridges were favoured human habitation sites throughout the Holocene. They were (and are) relatively mosquito-free, did not become sticky quagmires during the wet season, and supported a denser cover of tall shady trees than the adjacent, often marshy alluvial clay flats.

7.4.6 *Jebels*

In addition to congregating on top of sand ridges overlooking the White Nile and its littoral swamps or maya'as, the early inhabitants of the Gezira also showed a preference for rocky hills or jebels, a preference not shared by the peasant farmers of today. Where the jebel was large, with permanent water, and with extensive areas of level, stone-free soil, as at Jebel Moya (Figure 7), settlement was dense. At Jebel Tomat, which consists of two low granitic hills separated by a low flat col, occupation was on the col. The hill sites had several advantages, one of which was a supply of fresh stone suitable for implements, as described by Vail in chapter 2. In a land of sand and clay, where rocky hills were few and far from the fishing grounds, the early inhabitants of the Blue and White Nile valleys would naturally have made full use of wood, bone and shell. As gathering of wild cereal grasses became superseded by deliberate cultivation of millet and sorghum, and the need arose to store grain in an accessible but rodent-proof container, early potters from

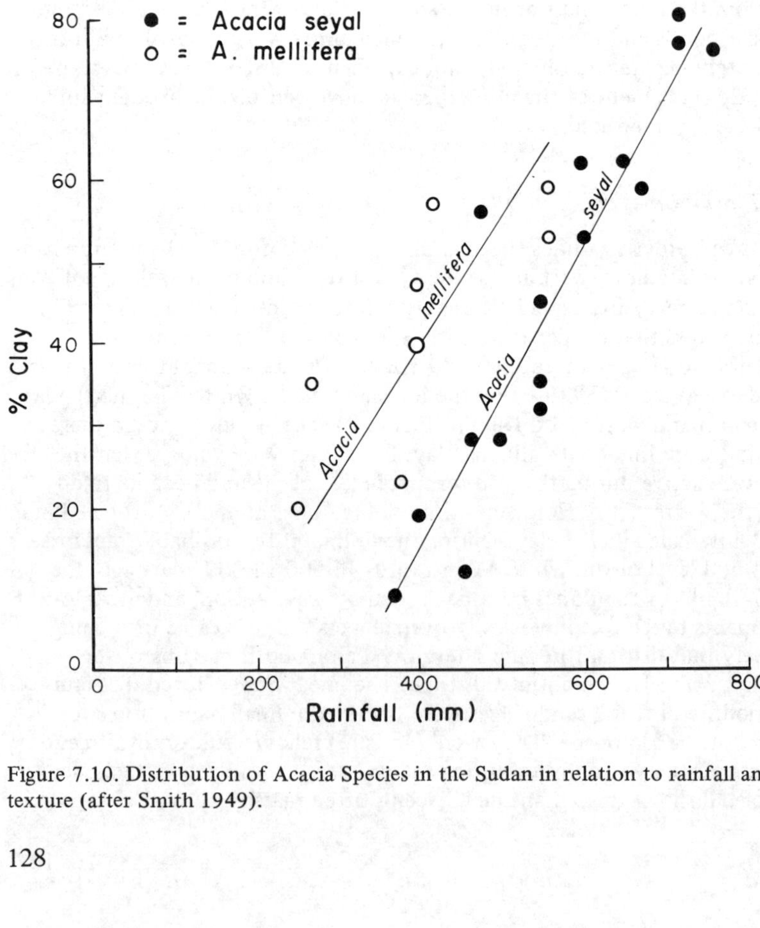

Figure 7.10. Distribution of Acacia Species in the Sudan in relation to rainfall and soil texture (after Smith 1949).

the dune sites would have had access to ample clay and sand. Despite the abundance of sand suitable for temper, the potters evolved an ingenious technology involving the use of freshwater sponge spicules as a temper for their pots. Adamson's discovery of such pottery at Jebel Tomat in 1972 suggests that seasonal movement between jebel and White Nile was in evidence some 1 500-2 000 years ago, if not earlier. A similar seasonal migration of the semi-nomadic inhabitants of the southern Gezira to and from the White Nile persists to this day.

7.5 SOILS AND LAND-USE

7.5.1 *Soils and vegetation*

In discussing some of the advantages of dunes over clay plains as prehistoric settlement sites, we mentioned their greater abundance of tall shade trees. Although clay soils absorb and retain more water than do sands, much of that water is not available for plant growth. Smith's now classic study of the distribution of tree species in the Sudan in relation to rainfall and soil texture demonstrated that a given species of Acacia growing on sand requires only two-thirds of the annual rainfall needed by that same species when growing on clay (Smith 1949 and Figure 10). During the 30 year period 1931-1960 rainfall in the Sudan ranged from over 1 400 mm in the south to less than 10 mm in the north (Figure 11). Since this range far exceeds the likely span of Holocene rainfall fluctuations in the Gezira discussed in Chapter 9, the relationship deduced by Smith probably also held good in central Sudan during the past 11 000 or so years. The present-day distribution of grasses, shrubs and trees in central Sudan is the subject of the next chapter, and provides a useful starting point from which to deduce what climatically-induced shifts in vegetation may have taken place during prehistoric times. Our concern here is with the distribution of major soil types which, with rainfall, control plant distribution in central Sudan.

7.5.2 *Distribution of soils*

Apart from the sedentary soils of the clay uplands shown on Figure 7, the soils in the Gezira are generally of Holocene age (Williams 1966, Abedine et al. 1971, Williams & Adamson 1976). There is a close correspondence between soil type and parent sediment, as shown in Figures 7, 9 and 12. Figure 12 is based upon a reconnaissance soil survey along the right bank of the White Nile by Williams during 1963-64 (Hunting Technical Services 1964), and shows that there is considerable variation in the physical and chemical properties of the clay soils bordering the White Nile. This variation is reflected in the vegetation cover, and is discussed in the next chapter.

Except at the northern end of the Gezira fan, the alluvial clays laid down by the Blue Nile are less variable, and coincide with the main Gezira irrigation area north and east of the Managil Ridge (Figures 1 and 7). The typical

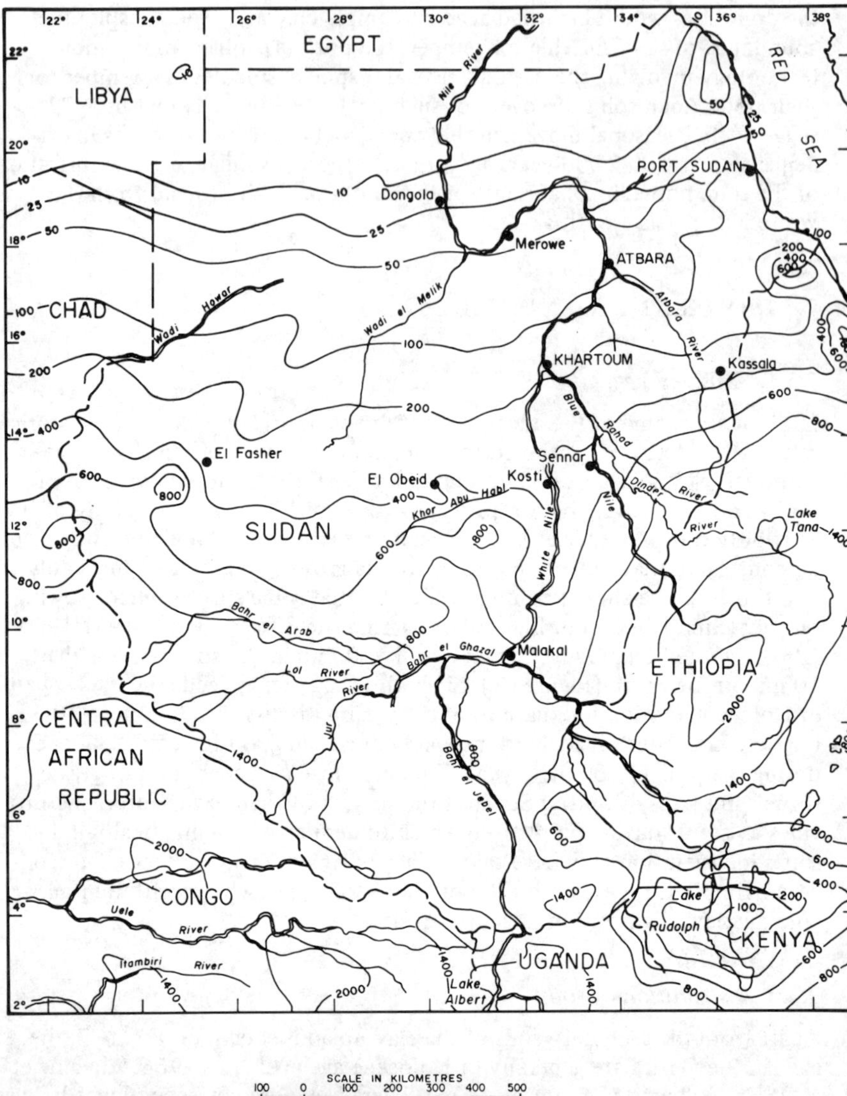

Figure 7.11. Average annual rainfall in Sudan (1931-1960) (after Sudan Meteorological Service).

Gezira cotton soil is a dark cracking clay (or vertisol) which is sticky and impermeable when wet, and fissured with a polygonal network of coarse and fine shrinkage cracks when dry (Finck 1961, Abedine & Robinson 1971, Abedine et al. 1969 and 1971). Owing to its economic significance, the

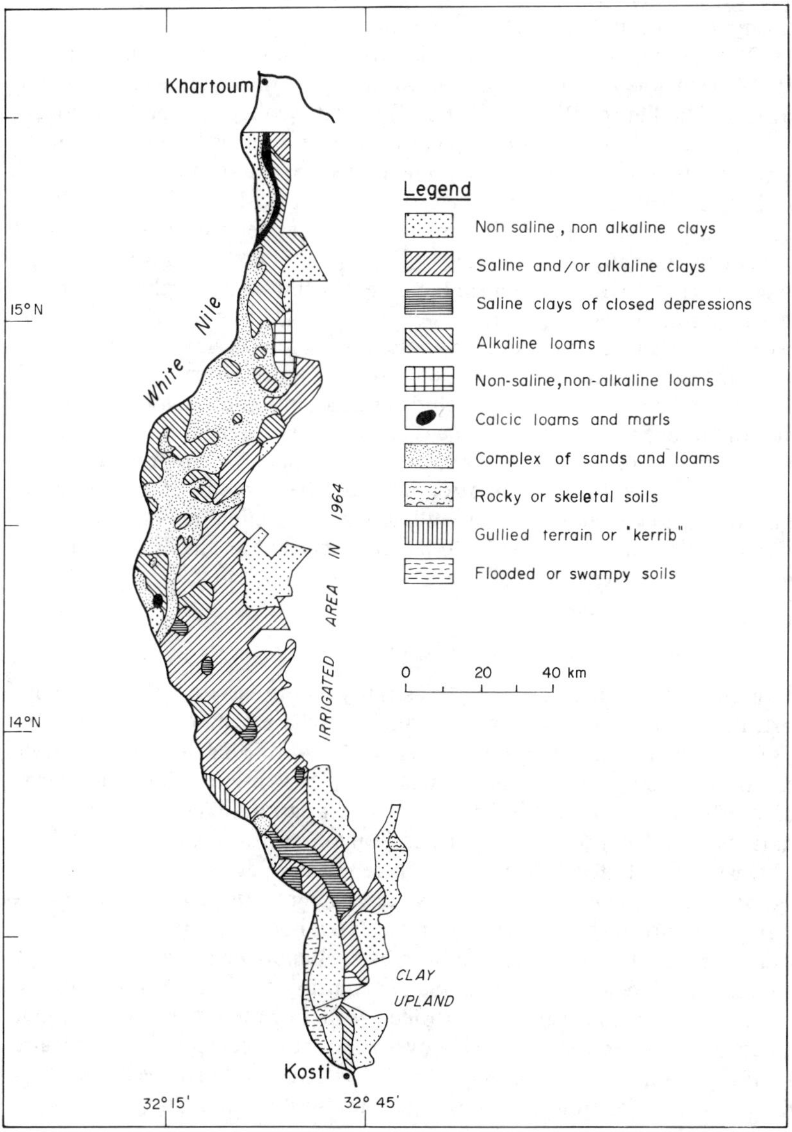

Figure 7.12. Soils of the western Gezira (adapted from Hunting Technical Services 1964; and Williams 1968b).

Gezira clay has been the subject of detailed chemical and pedological study (Greene 1928, Jewitt 1955, De Vos & Virgo 1969, Blockhuis et al. 1970, Bursink 1971, Fadl 1971, Abdulla 1972-73, Mukhtar et al. 1974).

An important present-day concern, and one which is also highly relevant to attempts to reconstruct the changes in vegetation in central Sudan (Wickens 1975 and chapter 3), is whether the Gezira clay is becoming progressively more saline and alkaline through time.

Half a century of sustained irrigation using Blue Nile water has so far added a negligible amount of sodium to the upper metre of Gezira clay (Jewitt & Middleton 1955, Rai 1965). The high quality of Blue Nile irrigation water, the absence of a rising water-table, and the very slow pace of exchangeable sodium accumulation augur well for the future: salinity and alkalinity hazards are not expected in the eastern Gezira until the middle of next century. Since the growth of many plants besides cotton is impeded by an excess of exchangeable sodium (USDA 1954, Finck 1963, Robinson 1971), the lack of any obvious long-term increase in salinity and alkalinity in the eastern Gezira during this century is circumstantial evidence that there was probably no significant increase in soil salinity and alkalinity in that area during the time of recurrent Holocene flooding by the Blue Nile and its distributaries. If so, we can safely exclude changes in soil chemistry as a major cause of Holocene vegetation changes in the eastern Gezira. Key factors responsible for changes in the plant cover could have been a progressive reduction in the extent and duration of Blue Nile flooding, accentuated by an unfortunate coincidence between late Holocene climatic desiccation and growing biological pressure exerted by the expansion of man's pastoral and agricultural activities in this region.

7.5.3 *Soil salinity along the White Nile*

The argument outlined above applies strictly to the eastern Gezira. In the west, the prognosis is less optimistic, and there have been recurrent phases of salt accumulation in the soils bordering the White Nile. The major factors that limit cotton yields in central Sudan are clay content, salinity and alkalinity. Where salinity and alkalinity are low, variations in cotton yield are significantly related to clay, organic carbon and nitrogen content (Finck & Ochtman 1961, Finck 1963). Although Jewitt & Middleton (1955) and Rai (1965) found no evidence of excessive salt accumulation or soil deterioration in irrigated cotton soils in the eastern Gezira, a number of irrigation blocks in the western Managil Extension have been abandoned since 1962 owing to rising salinity levels and declining yields (Hunting Technical Services 1964b). Part of the problem lies in the stratified profiles of the western soils, which consist of an upper horizon of clay over permeable silts and loams. Irrigation of alluvial plains with permeable subsoils has caused difficulties elsewhere, with seepage losses from canals, rising ground-water, increasing salinization, and lowered crop yields.

The mapped pattern of surface and subsurface salinity east of the White Nile (Williams 1968b) shows that there is some relationship between the inferred position of the early Holocene White Nile strandline and associated embayments and the presence of localized zones of high subsoil salinity (Figure 13). Whether the salts were initially derived from the underlying late Pleistocene evaporite deposits which border the White Nile at shallow depth (Williams & Adamson 1974) is not known. Evaporation around the margin

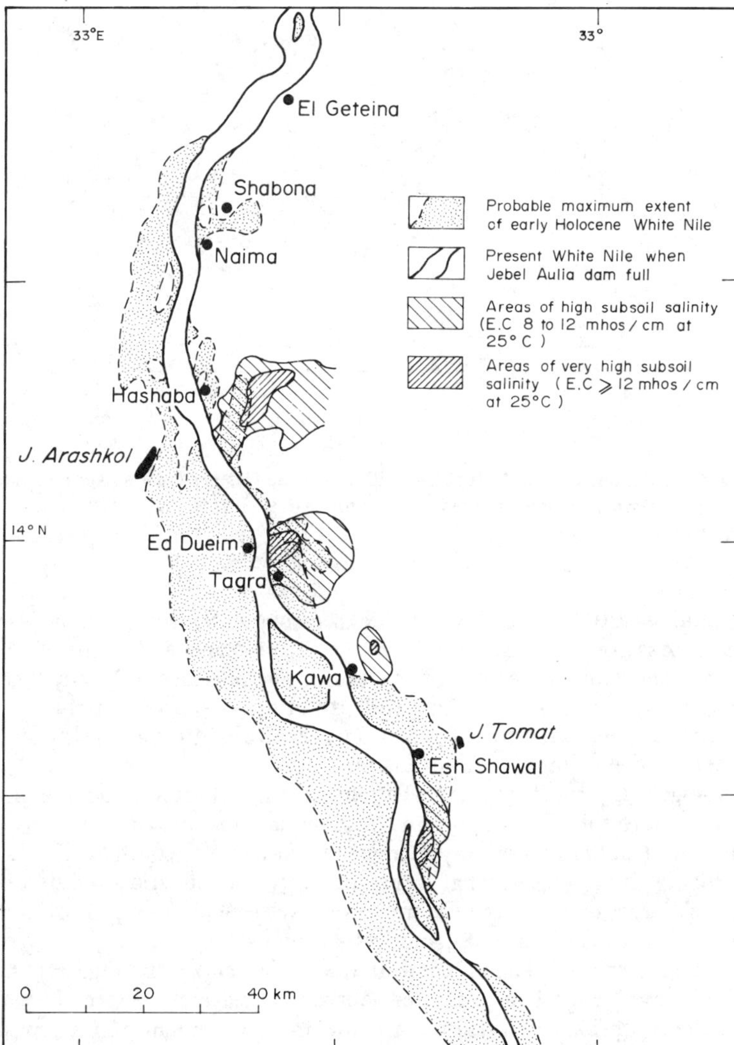

Figure 7.13. Subsoil salinity in the western Gezira in relation to the probable early Holocene extent of the White Nile (adapted from Williams 1968b; and Williams & Adamson 1976).

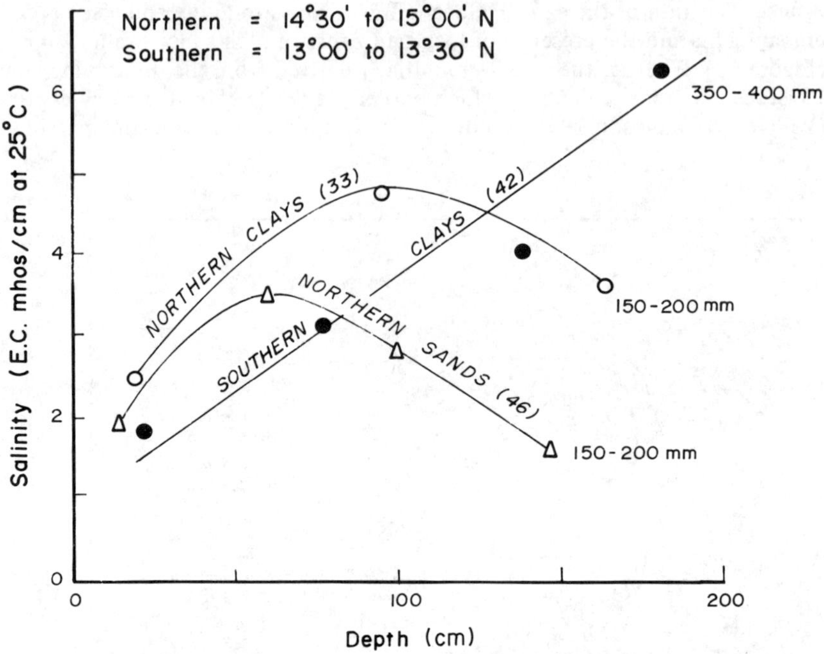

Figure 7.14. Relation between soil depth and salinity in the western Gezira. Numbers in brackets refer to number of sites sampled (after Williams 1968b). Mean rainfall shown for north and south.

of the dwindling early Holocene White Nile (see chapter 9) would in any case have caused some concentration of salts in the littoral zone, a phenomenon common in other semi-arid regions of the world. The presence of locally high concentrations of soluble salts at shallow depth in the western Gezira is important, for unwise extension of irrigation agriculture into these areas can and may aggravate existing salinity problems.

At this stage it is pertinent to ask whether present wet season leaching of salt from topsoil to subsoil is important. In sandy and loamy soils between Jebel Aulia and Esh Shawal, salinity increases from north to south at all depths, which is the reverse of what we might expect on climatic grounds. The south of the area depicted in Figure 13 receives roughly twice as much rain as the north (Figure 15), and wet season evaporation is four times higher at Khartoum than at Kosti. In the case of clay soils (Figure 14), the southern clays are leached to greater depths than clays in the semi-arid north. Northern topsoils are usually non-saline, but salt content rises sharply in the second horizon, with a gradual decrease thereafter (Figure 14).

Topography also influences soil salinity. Run-on sites are usually saline, and run-off sites are non-saline. Run-on sites include the clay-floored swales between dunes, and depressions on the clay plain. The extensive closed depressions east of Hashaba are saline at all depths (Figures 12 and 13). Many of these depressions represent former maya'as or swamps which became isolated from the White Nile during the Holocene. The effects of poor surface drainage, combined with an initial concentration of salts in the swamp sediments, has produced soils which support a sparse cover of stunted *Acacia nubica* in contrast to the dense thickets of *Acacia mellifera* on the well-drained upland clays further south. The flat swales between sand ridges are also run-on sites, and are generally very saline and devoid of trees. East of El Geteina, lacustrine marls occur in some of the larger swales. The calcareous soils of these former lake and swamp floors are highly saline and alkaline (Williams 1968a, b).

A detailed analysis of the factors influencing salt distribution in the western Gezira would be out of place here, but enough has been said to show that dominant salinity controls are soil texture, permeability and topography, all of which are related to depositional history. Not only has the physical landscape been evolving throughout man's occupation of the White Nile valley, but the more subtle control exerted by soil chemistry upon the flora, and so indirectly upon the fauna, has also been in a state of flux. The essential point to retain is that the present salinity pattern in the White Nile valley is not a unique function of present-day rainfall distribution, but is intimately bound up with the changing history of the White Nile.

7.5.4 *Ground-water and salinity*

In describing the landforms and soils of the Gezira in relation to the past and present inhabitants of this region we have so far focussed our attention upon such matters as shade and shelter, access to raw materials for pottery and stone implements, and proximity to the banks and maya'as along the White Nile, with their abundance of fish, waterfowl, molluscs, and riverain or riparian mammals. Equally vital to man's survival when away from the two Niles is a reliable supply of drinking-water. Although wells up to ten or so metres deep can certainly be dug with stone-weighted digging sticks, digging deep wells through the clays, sands and gravels of the Gezira fan in order to tap the aquifers of the Nubian Sandstone and of the Gezira Formation, difficult enough today for villagers lacking drilling equipment, was no doubt beyond the immediate requirements and capabilities of early man. The two most likely sources of freshwater in the early Holocene central Gezira during the dry season would have been seeps or springs along the lower slopes of the southern jebels, and the sandy distributary channels further north. Shallow wells 2 to 3 m deep are dug today in the beds of ephemeral or seasonal stream channels by the nomads of the Red Sea Hills, and of Kassala, Kordofan and Darfur Provinces, as well as by the Tuareg of Niger and the Bedu of Jordan.

Shallow wells dug within 5-10 km of the White Nile will tap ground-water, and during the early Holocene, before the Blue Nile cut down through its flood plain, the water-table near the Blue Nile would also have been near the surface. Ground-water near the Blue Nile is non-saline (300-500 ppm), as is ground-water near the White Nile (1 000 ppm), but it rapidly becomes more saline away from the two rivers, attaining values of 4 200 ppm immediately north of the Manaqil Ridge near El Azazi, and values of 32 000 ppm north of Abu Quta in the centre of the Gezira (chapter 5). Maximum salinity values coincide with a deep infilled basin in the central Gezira described in chapter 5. Boushi and Abdel Salam have proposed two possible reasons for the high ground-water salinity in this area:

1. The slowly moving ground-water is progressively dissolving the soluble salts from the sediments through which it is flowing, and

2. The prior accumulation of saline ground-water is associated with the deposition of evaporites within closed basins during drier intervals in the past.

Given the present state of our knowledge, we cannot choose between either hypothesis. The occurrence of localized evaporite deposits (gypsum, carbonate) within the Gezira Formation is not in doubt, but their precise extent and thickness is at present unknown. If the White Nile is extensively flanked

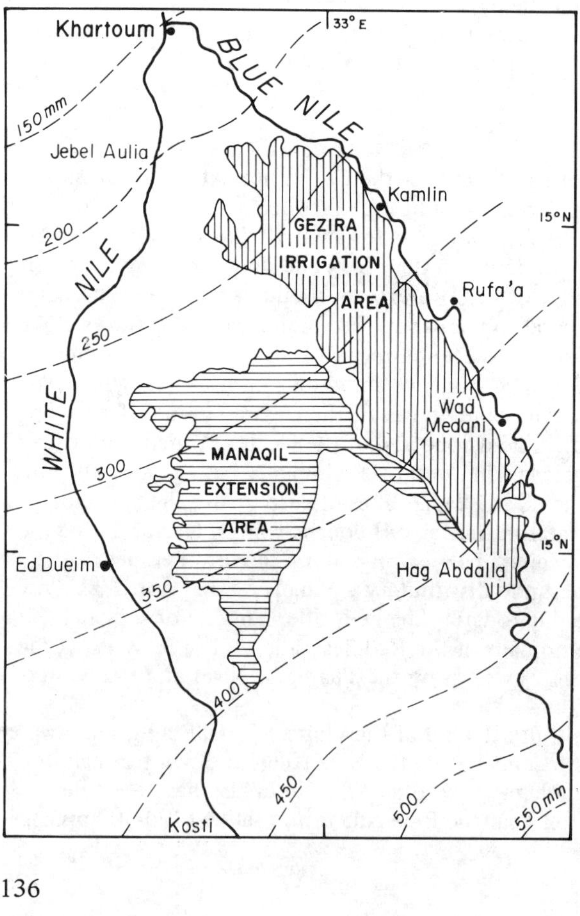

Figure 7.15. Mean annual rainfall in the Gezira 1921-1950 (after Randell 1961).

by highly saline late Pleistocene evaporites, which seems quite possible (Williams & Adamson 1974 and chapter 9), then any fresh water flowing out from the White Nile eastwards into the Gezira or westwards into Kordofan will rapidly become saline after passing through these evaporites. In point of fact, ground-water salinity values do indeed rise very rapidly away from the White Nile, so that even if the early inhabitants of the Gezira had tried to dig deep wells away from the river, their efforts would have been poorly rewarded.

7.6 CLIMATE: THE PROBLEM OF WATER

7.6.1 *Rainfall, temperature and evaporation*

In all semi-arid areas, the chief pre-occupation of man and animals is water. Without water, plants die, and animals must migrate, or die. On average, the Sudanese peasant farmer dependent on raingrown dura (sorghum) or dukhn (millet) expects one bad year in three. During 1968-73 the succession of dry years caused considerable hardship, and widespread destruction of the vegetation throughout the semi-desert scrub region of the Sudan. The Gezira lies within this region, and has some eight months of hot dry weather during which the dominant winds blow from the arid north-east, followed by the harif or rainy season of three to four months during which the dominant winds are from the south-west. Average annual rainfall decreases steadily to the north, as shown in Figures 11 and 15. Kosti (406 mm) receives over twice as much rain as Khartoum (164 mm), has a longer growing season and experiences greater intensity of rainfall (Hunting Technical Services 1964). The two rainfall maps show that the isohyets swing away north-east from Kosti, so that localities in the White Nile valley receive less rain than places of similar latitude in the eastern Gezira. The sand deposits along the White Nile compensate for this lack of rainfall, and until quite recently were thickly wooded (chapter 8).

Figure 16 shows the mean maximum and minimum monthly temperatures at five representative stations in central Sudan. Highest temperatures are in May at all stations except Kosti, where April is the hottest month, owing to its more southerly latitude and earlier onset of the rains. There is a close relationship between mean daily maximum temperature and mean daily evaporation, as indicated in Figure 1 of chapter 8. The southern stations have the lowest evaporation rates during all but the winter months, when evaporation in the cooler north is at a minimum.

7.6.2 *Water and plant growth*

The growing season begins earlier and ends later in the south than in the north. When 50 mm of rain a month are falling at Kosti, Khartoum receives less than 10 mm; in July Kosti receives over 120 mm of rain, and Khartoum less than 50 mm; in August over 140 mm fall at Kosti, but only 70 mm at Khartoum. If the effects of higher summer evaporation at Khartoum are also

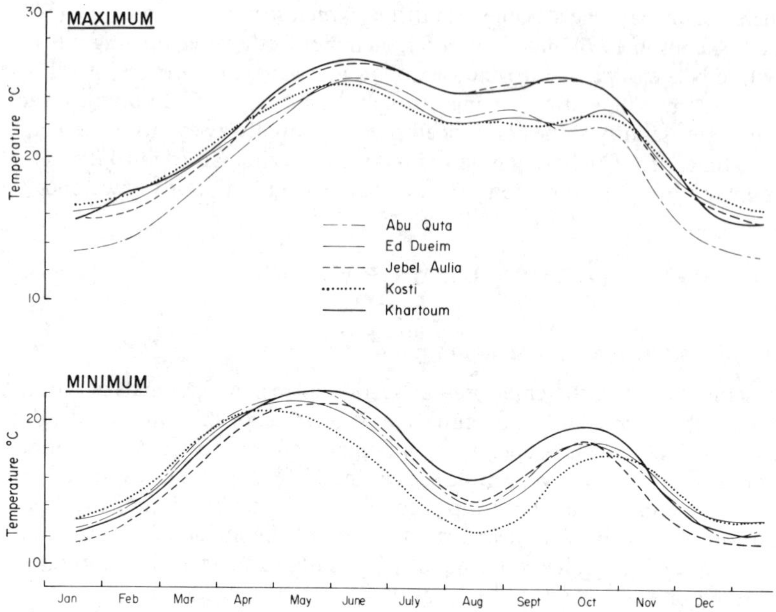

Figure 7.16. Mean maximum and minimum monthly temperatures at five stations in central Sudan (after Sudan Meteorological Service and Hunting Technical Services 1964b).

considered, then it becomes clear why Kosti has a growing season roughly twice as long as Khartoum; extending from May-June until late September or early October.

We will argue at greater length in chapter 9 that the early Holocene climate in the northern Gezira was akin to that of Kosti today, and that Kosti probably had a climate comparable to that now experienced at Malakal. We will also suggest that the extent of wet season flooding by both the Blue Nile and the White was far greater during the first half of the Holocene than during the past 5 000 or so years. The significance of these environmental changes for early man was summarised in June 1973 in a preliminary archaeological report by Professor J.D. Clark:

'No evidence of domestic animals or plants was recovered and it may be inferred that the Shabona sand ridge site in the lower White Nile valley was occupied by a group of hunter/fishers and gatherers; it is likely to have been seasonally rather than permanently occupied. The quantities of grindstones (about the commonest artifact) suggest that wild grass seeds were an important source of food. Fish also formed an important part of the diet. The most favourable time of the year for the site to have been occupied, therefore, would have been when the floods were beginning to subside and many lagoons and backwaters were stocked with fish – from November onwards. Today local rains fall mainly in July and August but, if the rainfall at the time the site was occupied was more comparable to that recorded further south, where the rainy season spreads over seven months, then occupation in November, as suggested, can be expected to have coincided also with the time when a number of wild cereal grasses were ready for harvesting and when probably also the game would begin to return to the main river.'

Evidence from sub-fossil mollusca in lake and river deposits both in and away from the Blue and White Nile valleys is fully consistent with such a reconstruction. In 1944 Tothill noted the presence of the amphibious mollusca *Pila wernei* and *Lanistes carinatus* throughout the upper 180 cm of the Gezira clay plain. He concluded that 'the regular vertical distribution at all levels' of these semi-aquatic snails 'almost as far north as Khartoum shows that the climate at the time of deposition was wet enough to enable these species to survive the dry season' as they do today near Malakal. The argument is not entirely convincing, for the early Holocene floodplain of the Blue Nile would in any case have been far swampier, regardless of the local rainfall at that time. More persuasive is the presence of dense populations of the land snail *Limicolaria flammata* at least as far north as Khartoum towards the end of the interval of widespread early to middle Holocene Blue Nile flooding (Tothill 1944, 1946). These snails are very rare today north of Wad Medani, are uncommon near Sennar, and only become abundant in the Acacia Tall Grass country around Gedaref, where present-day rainfall is over 500 mm a year.

Limicolaria and *Zootecus* shells were recovered from the top few centimetres of now dry calcareous pans west of Jebel Aulia which contained permanent water during one or more intervals between 8 500 and 7 000 years ago (Williams et al. 1974). The land snails suggest higher rainfall near Khartoum at this time for they were washed into small lakes which supported an abundant fauna of permanent water mollusca, and which were totally dependent on local rainfall and runoff for their continued existence.

7.6.3 *Impact of drought: past and present*

In two elegantly presented and persuasively argued archaeological monographs, both of which are now classics, Arkell (1949, 1953) has marshalled detailed evidence showing Nile flood levels and wetter climates at and north of Khartoum during early and middle Holocene times some 8 000 to 5 000 years ago. During the time of the 'Early Khartoum' culture, the country around Khartoum seems to have consisted of swamps and grassy plains 'such as those on which the Nile Lechwe flourishes in the southern Sudan today' (Arkell 1953:9). By Neolithic times, the northern Gezira and the Nile valley immediately north of Khartoum were probably a partly wooded steppe, unlike the almost treeless semi-desert of today (Arkell 1953:9). The presence of *Celtis* seeds and of *Limicolaria flammata* shells 'speaks for an average rainfall of at least 500 mm in Neolithic times, with a probably greater rainfall in Mesolithic times' (i.e. Early Khartoum times, c. 8 000 BP, Adamson et al 1974) 'for where the Nile Lechwe flourishes today the average rainfall is 800-900 mm' (Arkell 1953:9).

Arkell's suggestion that climate was wetter at Khartoum during the early Holocene was accepted by Tothill with this qualifying remark:

'The adult semi-fossil shells of *Limicolaria* obtained by A.J. Arkell at his Hospital site excavation are notably smaller than is normal today for Suki to Gedaref specimens, which suggests that the climate was changing to the present one in which this conspicuous cone-shaped land mollusc is uncommon north of Singa and unknown except as a semi-fossil north of Wad Medani' (Tothill 1948:139).

It is an oversimplification to think of a progressive climatic deterioration in the Sahel and Savanna zones of Africa during the Holocene (chapter 2). There were at least four Holocene intervals of high lake levels in Ethiopia, Chad and Niger, but it is now certain that the early Holocene climate in these regions was wetter than during the past 3 000-4 000 years (Servant 1974; Gasse 1975; Williams et al. 1977). Put succinctly, the early Holocene was wet with drier intervals, and the late Holocene was dry with moister intervals. No doubt the southward shift of rainfall, vegetation and faunal zones accelerated the trend towards agriculture and pastoralism, a trend which may itself have aggravated the biological consequences of the deteriorating late Holocene climate. In chapter 3 we have explored some of the consequences of the most recent severe drought upon the flora of the Sudan, and in chapter 2 we have examined the late Quaternary climatic fluctuations throughout the Sudan and Sahel zones of northern Africa. Detailed discussion of the late Quaternary environments of the Gezira is the subject of chapter 9.

An important conclusion to emerge from the present chapter is that the physical landscape of the central Sudan is no more static than the organisms that dwell upon it: the present does indeed furnish a key to the past, but the past is also a very real key to the present.

7.7 REFERENCES

Abdulla, H.H. 1972-73, Profile studies in the Shawal-Manaqil Extension Area. Ann. Rept. Gezira Res. Station and Sub-stations.

Abedine, A.Z. el & G.H. Robinson 1971, A study on cracking in some vertisols of the Sudan. *Geoderma* 5:229-242.

Abedine, A.Z. el, G.H. Robinson & J. Tyego 1969, A study of certain physical properties of a vertisol in the Gezira area, Republic of Sudan. *Soil Sci.* 108:359-366.

Abedine, A.Z. el, G.H. Robinson & A. Commissaris 1971, Approximate age of the vertisols of Gezira, central clay plain, Sudan. *Soil Sci.* 111:200-207.

Adamson, D., J.D. Clark & M.A.J. Williams 1974, Barbed bone points from central Sudan and the age of the 'Early Khartoum' tradition. *Nature* 249:120-123.

Addison, F. 1949, *The Wellcome excavations in the Sudan. Jebel Moya.* Oxford University Press.

Andrew, G. 1944, Notes on Quaternary climates in the Sudan. Appendix XXV in Soil Conservation Committee's Report, Sudan Govt, Khartoum.

Arkell, A.J. 1949, *Early Khartoum.* Oxford University Press.

Arkell, A.J. 1953, *Shaheinab.* Oxford University Press.

Badri, Omer el 1972, Sediment transport and deposition in the Blue Nile at Khartoum, flood seasons 1967, 1968 and 1969. M.Sc. thesis, Univ. Khartoum.
Baker, Sir Samuel 1866, *The Albert N'Yanza. Great basin of the Nile and explorations of the Nile sources.* Reprinted 1962. London, Sidgwick & Jackson.
Balfour Paul, H.G. 1952, Early cultures on the northern Blue Nile. *Sudan Notes and Records* 33:202-214.
Bishai, H.M. 1962, The water characteristics of the Nile in the Sudan with a note on the effect of *Eichhornia crassipes* on the hydrobiology of the Nile. *Hydrobiol.* 19:357-382.
Blockhuis, W.A., S. Slager & R.H. van Schagen 1970, Plasmic fabrics of two Sudan vertisols. *Geoderma* 4:127-138.
Bunting, A.H. & J.D. Lea 1962, The soils and vegetation of the Fung, east central Sudan. *J. Ecol.* 50:529-558.
Buursink, J. 1971, *Soils of Central Sudan.* Utrecht.
Clark, J.D. 1973, Preliminary report of an archaeological and geomorphological survey in the central Sudan, January to March, 1973. Unpublished report, June, 1973, lodged with Dir. of Antiquities, Khartoum.
De Vos t. N.C., J.H. & K.J. Virgo 1969, Soil structure in vertisols of the Blue Nile clay plains, Sudan. *J. Soil Sci.* 20:189-206.
Fadl, A.E. 1971, A mineralogical characterization of some vertisols in the Gezira and the Kenana clay plains of the Sudan. *J. Soil Sci.* 22:129-135.
Finck, A. 1961, Classification of Gezira clay soil. *Soil Sci.* 92:263-267.
Finck, A. 1963, *Tropische Böden.* Hamburg, Parey.
Finck, A. & L.H.J. Ochtman 1961, Problems of soil evaluation in the Sudan. *J. Soil Sci.* 12:87-95.
Gaitskell, A. 1959, *Gezira: a story of development in the Sudan.* London, Faber & Faber.
Gasse, F. 1975, L'évolution des lacs de l'Afar central (Ethiopie et TFAI) du Plio-Pléistocène à l'Actuel: reconstitution des paléomilieux lacustres à partir de l'étude des Diatomées. D.Sc. thesis, 3 vols, Univ. Paris:406.
Greene, H. 1928, Soil profile in Eastern Gezira. *J. Agric. Sci.* 18:518-530.
Hunting Technical Services Ltd 1964, Report 6, Roseires soil survey project. *White Nile right bank Khartoum to Rabak, soils and engineering reconnaissance.*
Hunting Technical Services Ltd 1966, Report 12, vol. 3, Roseires soil survey project. Blue Nile west bank Abu Hugar to Sennar.
Hurst, H.E. & P. Phillips 1931, *The Nile basin. General description of the basin.* Cairo, Phys. Dept. Paper, Govt. Press.
Jewitt, T.N. 1955, *Gezira soil.* Republ. of Sudan, Min. Agr. Bull. 12, Khartoum.
Jewitt, T.N. & K.R. Middleton 1955, Changes in Sudan Gezira soil under irrigation. *J. Agric. Sci.* 45:277-279.
Mukherjee, R., C.R. Rao & J.C. Trevor 1955, *The ancient inhabitants of Jebel Moya (Sudan).* Cambridge University Press.
Mukhtar, O.M.A., A.R. Swoboda & C.L. Godfrey 1974, The effect of sodium and calcium chlorides on structure stability of two vertisols: Gezira Clay from Sudan, Africa, and Houston Black Clay from Texas, USA. *Soil Sci.* 118:109-119.
National Council for Research 1975, *Agricultural Research Capabilities and Scope in the Sudan.* Agric. Res. Council, NCR, Khartoum.
Rai, K.O. 1965, Effects of long-term irrigation on Gezira soil. Ann. Rep. Gezira Res. Station and Sub-stations 1964-65 seasons.
Randell, J.R. 1961, A rainfall map of the Sudan Gezira. *Sudan Notes and Records* 42: 29-36.
Robinson, G.H. 1971, Exchangeable sodium and yields of cotton on certain clay soils of Sudan. *J. Soil Sci.* 22:328-335.

Servant, M. 1974, Les variations climatiques des régions intertropicales du continent africain depuis la fin du Pléistocène. Question 1, Rapport 8, 1-10, Société Hydrotechnique de France, XIIImes Journées de l'Hydraulique, Paris.

Smith, J. 1949, *Distribution of tree species in the Sudan in relation to rainfall and soil texture.* Min. Agric. Bull. 4, Khartoum.

Sudan Meteorological Service, n.d., Climatological normals (provisional) 1931-1960 (Khartoum).

Talling, J.F. 1957, The longitudinal succession of water characteristics in the White Nile. *Hydrobiol.* 11:73-89.

Talling, J.F. 1976, Water characteristics. In J. Rzóska (ed.), *The Nile. Biology of an ancient river.* The Hague, Junk:357-384.

Thompson, K. 1976, Swamp development in the head waters of the White Nile. In J. Rzóska (ed.), *The Nile. Biology of an ancient river:*177-196.

Tothill, J.D. 1944, A note on Darfur from a soil conservation point of view. Appendix XVIII in Soil Conservation Committee's Report, Sudan Govt., Khartoum.

Tothill, J.D. 1946, The origin of the Sudan Gezira clay plain. *Sudan Notes and Records.* 27:153-183.

Tothill, J.D. 1948, A note on the origins of the soils of the Sudan from the point of view of the man in the field. In J.D. Tothill (ed.), *Agriculture in the Sudan.* Oxford University Press:129-143.

US Department of Agriculture 1954, *The diagnosis and improvement of saline and alkali soils.* Handbook 60, Govt. Printing Office, Washington, DC.

Williams, M.A.J. 1966, Age of alluvial clays in the western Gezira, Republic of the Sudan. *Nature* 211:270-271.

Williams, M.A.J. 1968a, A dune catena on the clay plains of the west central Gezira, Republic of the Sudan. *J. Soil Sci.* 19:367-378.

Williams, M.A.J. 1968b, Soil salinity in the west central Gezira, Republic of the Sudan. *Soil Sci.* 105:451-464.

Williams, M.A.J. 1968c, The influence of salinity, alkalinity and clay content on the hydraulic conductivity of soils in the west central Gezira. *African Soils* 13:35-60.

Williams, M.A.J. & D.A. Adamson 1973, The physiography of the central Sudan. *Geogr. J.* 139:498-508.

Williams, M.A.J. & D.A. Adamson 1974, Late Pleistocene desiccation along the White Nile. *Nature* 248:584-586.

Williams, M.A.J. & D.A. Adamson 1976, *The origins of the soils between the Blue and White Nile rivers, central Sudan, with some agricultural and climatological implications.* Econ. and Soc. Res. Council, Natl. Council for Res. Sudan, Occasional Paper 6. Khartoum.

Williams, M.A.J., A.H. Medani, J.A. Talent & R. Mawson 1974, A note on Upper Quaternary mollusca west of Jebel Aulia. *Sudan Notes and Records* 54:168-172.

Williams, M.A.J., P.M. Bishop, F.M. Dakin & R. Gillespie 1977, Late Quaternary lake levels in southern Afar and the adjacent Ethiopian Rift. *Nature* 267:690-693.

M.OBEID MUBARAK
University of Gezira, Wad Medani

E.A.BARI
University of Khartoum

GERALD E.WICKENS
Royal Botanic Gardens, Kew,

MARTIN A.J.WILLIAMS
Macquarie University

8

The vegetation of the central Sudan

The character of the vegetation in the Sudan depends largely on rainfall and soil type, with the major divisions of the vegetation determined in the main by rainfall and the subdivisions by soil type (Andrews 1948; Smith 1949; Harrison & Jackson 1958; Bunting & Lea 1962). However, the biotic factor (grazing, cutting, burning and cultivation) is at least as important as the physical environment in determining the changing nature of this vegetation (Kassas 1956; Halwagy 1962a, b; Omer 1969; Obeid & Seif El Din 1970; Obeid & Mahmoud 1969, 1971).

The Gezira plain between the Blue and White Nile, Central Sudan, constitutes an area which has witnessed the Christian Kingdom of Soba (AD 548-1485) and the more recent 'Sultana El Zarga' or the Fung Kingdom (AD 1485-1820) which was based on Sennar. During its recent history of intensive irrigated crop production the Gezira has become one of the most heavily populated parts of the Sudan.

Under the present rainfall regime, agriculture in the form of the traditional shifting and 'harig' systems of cultivation may be assumed to have been practised for several millennia, especially in the central and southern parts of the Gezira. The essence of the shifting system is that the land is first cleared of its trees, bush and scrub, then during the rainy season dura, okra, legumes and some oil crops are grown. Weeding is by hoeing or pulling. Cultivation continues for some years, after which the soil loses its fertility and is left to revert to bush or grassland. Another area is cleared for cultivation and the cycle is repeated. In the harig system a 12-15 year old woodland or scrub with a dense understorey of grass is cut down in the dry season (usually January) and by the beginning of the first showers of the rainy season (usually June), when the new growing grass is showing green under the old one, the matted growth of grass is set alight. The heat produced in burning the dry grass is sufficient to kill the new growth and acts as a substitute for weeding. The same crops mentioned above are cultivated and, when soil fertility is lost, cultivators move on to another area.

The shifting system of cultivation is suited to the central and northern parts of the Gezira because tree or bush growth is sparse and the grasses are annual and not dense. The harig system of cultivation is practised in the

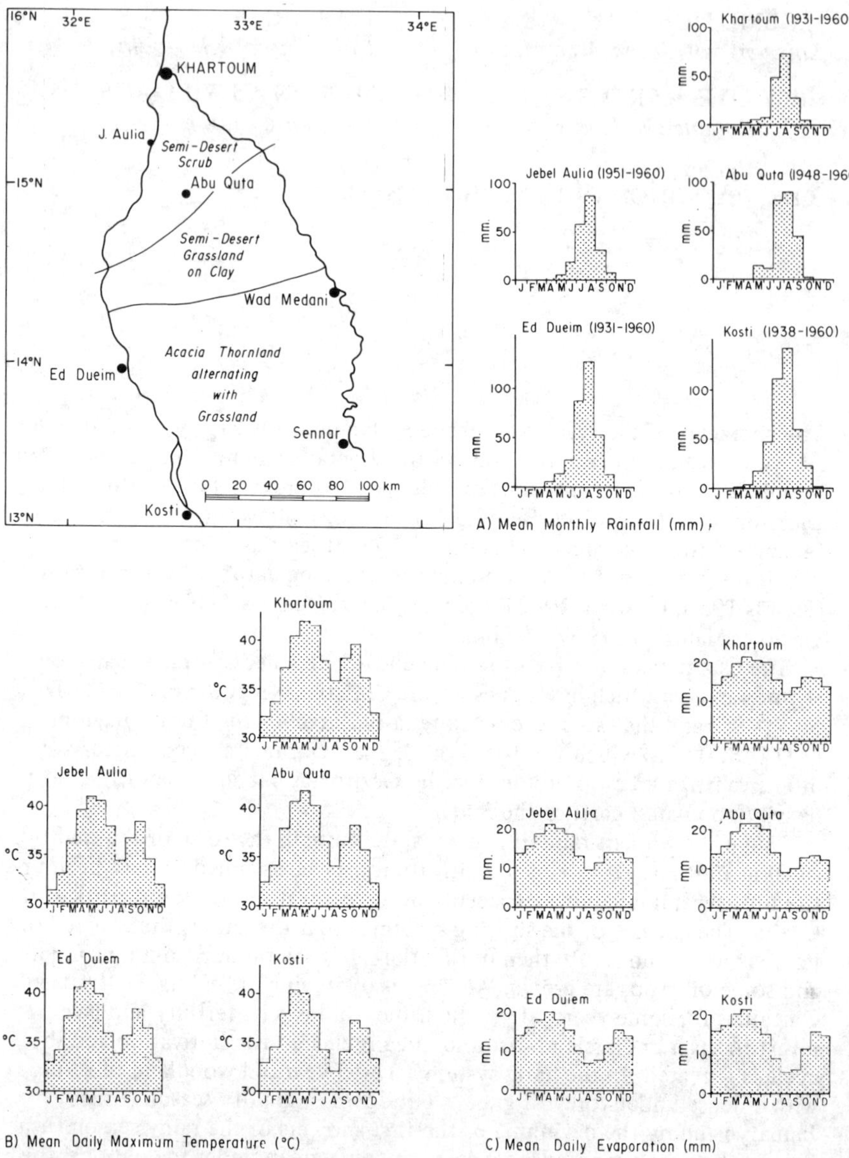

Figure 8.1. Generalized vegetation belts of central Sudan in relation to rainfall, temperature and evaporation. (Data from Sudan Meteorological Service and Hunting Technical Services 1964b.)

southern Gezira where the grasses are mainly perennial and both tree and grass growth are dense.

Throughout the recent history of the area the cultivation cycle seems to have remained basically the same, but there have been certain changes related to the continuously increasing population, the availability of drinking water for the increasing human and animal populations, and the establishment of irrigated agricultural schemes.

In areas near water supplies the increasing density of population has led to a lengthening of the period of cultivation and to a shortening of the bush fallow to three to six years. Where the water supply (for animal and human consumption) has seriously diminished, villages move to new sources. The abandoned land is eventually colonized by a series of secondary communities.

Where agricultural schemes have been established, the axe and the tractor were extensively used to clear scrub and woodland in order to use the land for growing crops. The Gezira Irrigation Scheme which was started on some 600 acres of land in 1911 at Tayiba (Gaitskell 1959) now covers two million acres of land on which are grown cotton, dura, groundnuts, wheat and legumes.

In addition to clearing woodland and scrub for agriculture and for providing firewood, the area has been greatly influenced by the grazing of cattle, sheep, camels and goats.

Although the area under consideration falls broadly within the three major vegetation units outlined by Harrison & Jackson (1958) in relation to rainfall as semi-desert scrub, semi-desert grassland on clay and *Acacia mellifera* thornland alternating with grass areas (Figure 1), topography, soil type, and the biotic factor (animal and human) seem to be major interacting factors affecting the distribution and floristic composition of the plant communities within the different units. We will first discuss the communities characteristic of each of these three broad units before going on to consider three 'azonal' associations in more detail. These are the swamp associations of the Sudd and White Nile floodplain; the riverine woodland associations of the several Niles; and the vegetation now found on the jebels.

8.1 SEMI-DESERT SCRUB

This community forms the northernmost part of the plain. Its mean annual rainfall varies from about 165 to about 300 mm. The bulk of this rainfall, which is very variable and unreliable, is confined almost entirely to the three months of July, August and September and comes as a few local storms or scattered showers (Oliver 1965; El-Tom 1972). In this belt, the coolest month (January) has a mean minimum temperature of about 15 °C and a mean maximum temperature of about 32 °C, and the hottest month (May) has a mean minimum temperature of about 25 °C and a mean maximum temperature of about 41 °C (Figure 1). These relatively high temperatures result in a high rate of evaporation, and, from the ecological point of view, the combination of low unreliable rainfall and high potential water loss are the most significant climatic variables. However, quantitatively, the available data often fall short of what is required to provide the best correlations between the distribution of plant communities and the climatological data available.

The soils can be classified as:

1. Riverain soils which are associated with the present flood plain of the Nile river system and include:

– White Nile clays: these soils contain 60-70 % clay and are rather uniform in texture, probably on account of their very slow deposition by the White Nile.

– Blue Nile silts and silty clays: in these the levee soils which lie next to the river tend to be light-textured and the lower soils which lie further away to be heavy-textured. However, these soils are rather variable in texture, perhaps reflecting the variable nature of the flooding.

– Miscellaneous soils ranging in texture from sand to silt with admixtures of finer materials. These occupy relatively small areas.

All these riverain soils are alkaline, and the fine-textured ones are widely cultivated.

2. High level dark clays: these occur above the present flood level of the river between the Blue and White Nile. The soils are overlain by sand in the very north of this area. These soils have a high salt content and contain calcium carbonate and gypsum. The soils are generally grey, becoming browner southwards, and have affinities with the cotton soils of the Gezira.

The vegetation of this semi-desert scrub consists of a varying mixture of grasses and herbs either without any woody vegetation at all, or, more usually, with a variable scatter of bushes up to 4 m high interspersed with bare areas in the riverain soils. However, the vegetation is seldom homogeneous owing to severe surface erosion. Topography seems to play a major role in the distribution of the species. The ground is divided into eroded land adjacent to the Nile, hard-surfaced runoff sites, desert pavement, and runon sites such as runnels and drainage lines.

The vegetation is represented by a number of plant community types which can be recognized by their floristic composition and habitat. These community types seem to repeat themselves with reasonable similarity whenever the habitat is repeated. However, each community type *vis-a-vis* the type of habitat has its most characteristic dominant species, but its floristic composition includes a number of associates. Species differ in their ability to associate with others, for example *Salvadora persica* is confined to a single community type associated with sub-dominant *Capparis decidua* on the high-level dark clays north and south of Jebel Aulia. By contrast, *Acacia tortilis* subsp. *tortilis* and *raddiana, A.nubica, Capparis decidua, Aristida funiculata* and *Schoenefeldia gracilis* have a wide 'sociological tolerance' and are observed in most community types. They seem to have a wide sociological range: a phytosociological feature indicating that the concept of fidelity which is characteristic of the vegetation of wet climates is also applicable to the vegetation of this arid climate. The main differences between the plant growth of these communities is not primarily attributed to the variations in the relative abundance of the species but to differences resulting from their absence or presence. A generalised map of the vegetation of the Gezira plain is given as Figure 2. The vegetation on the White Nile bank and riverain plain is based upon field mapping by Williams during 1963-64 (Hunting Technical Services 1964), and upon detailed collecting by Bari during 1973.

Acacia tortilis subsp. *tortilis,* a bushy tree up to 3 m high, with its characteristic umbrella-shaped growth form is the only feature of the vegetation that is generally constant. It is usually distributed as an even scatter, with a greater concentration of bushes along drainage lines, and, under high rainfall, some dense, even stands occur (Harrison & Jackson 1958). The species has a wide range of tolerance: it may grow on erosion pavements where the habitat is rocky and the soils are shallow, or may occupy water runnels with differing water budgets and varying soil depths. The latter range from intermediate water runnels, with a floor covered with an admixture of gravel and sand to large runnels with sandy floors. The performance of the species (indicated by its density and the size of individual trees) differs according to the type of habitat. *Capparis decidua* is usually present in considerable amounts and in some places becomes abundant.

Acacia tortilis subsp. *raddiana* is locally abundant on sandy drainage lines, especially along the Blue Nile riverain soils (Figure 2). *Salvadora persica* (the toothbrush tree) covers an extensive belt in the high-level dark clays south of Khartoum, particularly towards the White Nile (Figure 2). This species exhibits two growth-forms. It may grow into a small tree 5-7 m high, and this is apparently its 'normal' form. However, under repeated grazing and/or cutting it may form tangled thickets covering mounds or low hillocks of sand and silt. This phenomenon is also observed in other tree species which grow in the area, such as *Capparis decidua* and *Ziziphus spina-christi.* The former species is found in sufficiently-watered low-lying clay plains or shallow wadis draining these clayey habitats; the latter species is associated with the banks of the two rivers and with light soils.

Acacia nubica and *Calotropis procera* are common in the vicinity of villages and are indicators of heavy grazing pressure. A combination of loamy soil and overgrazing seems strongly to influence the distribution of *Acacia nubica.*

Non-cracking but not hard-surfaced clayey areas are characterized by the presence of *Acacia ehrenbergiana* rather than *A.tortilis. Balanites aegyptiaca* is present along the clay drainage lines near the Blue Nile, whereas *Acacia nilotica, A.albida* and *A.seyal* are closely associated with the inundated areas along the banks. Even further south in the wetter regions, *Acacia nilotica* is never found far from standing water. It usually occurs in two types of habitat: on river banks where it is subjected to seasonal flooding for three to four months every year, and at the outlets of wadis which pour into the Nile and receive flood water. *Acacia seyal* is not normally a tree of arid and semi-arid country, but is one of the vegetation types of the Low Rainfall Woodland Savanna (Harrison & Jackson 1958) with a rainfall of over 450 mm

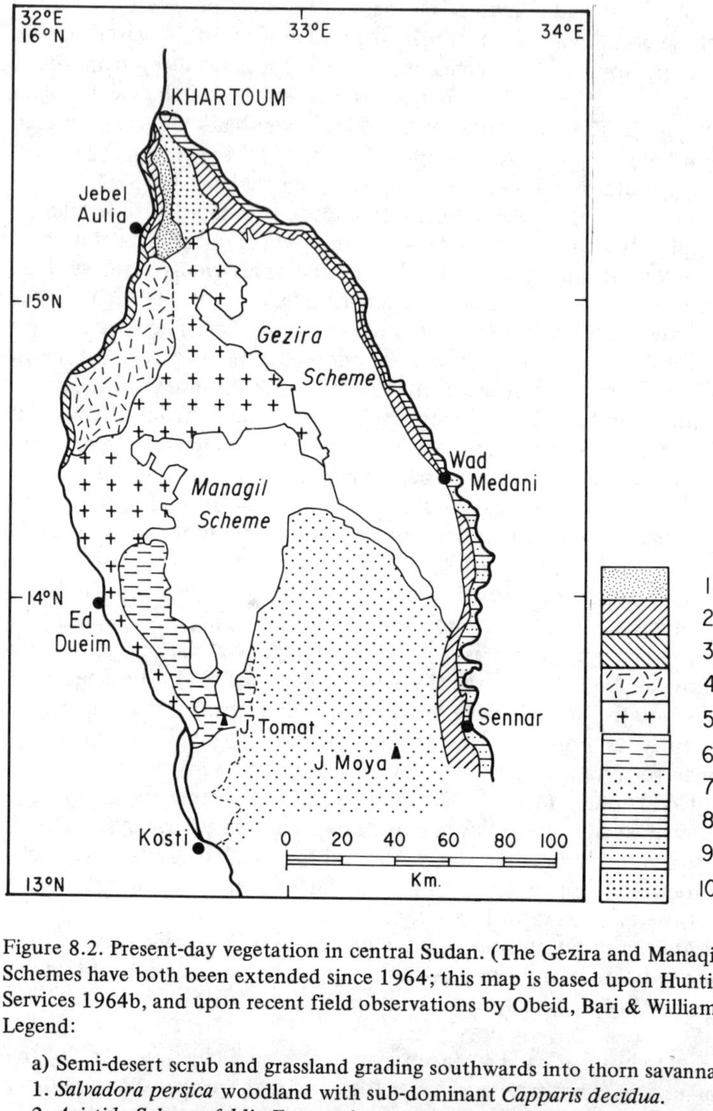

Figure 8.2. Present-day vegetation in central Sudan. (The Gezira and Manaqil Irrigation Schemes have both been extended since 1964; this map is based upon Hunting Technical Services 1964b, and upon recent field observations by Obeid, Bari & Williams.)
Legend:

a) Semi-desert scrub and grassland grading southwards into thorn savanna:
1. *Salvadora persica* woodland with sub-dominant *Capparis decidua.*
2. *Aristida-Schoenefeldia-Eragrostis* grasslands, with *Cassia senna* dominant near settlements in the north, and *Acacia nubica* in the south.
4. *Panicum turgidum* grassland with occasional *Capparis decidua.*
5. *Aristida* grassland with occasional *Acacia nubica* and *Capparis decidua.*
6. *Acacia nubica* scrub with ground flora of *Aristida* species and *Schoenefeldia gracilis.*
7. *Acacia mellifera-Acacia nubica* woodland alternating with tall *Cymbopogon* grassland.
10. *Aristida* grassland with *Acacia ehrenbergiana* and *A.tortilis.*

per year. However, in this semi-arid area the species is confined to sufficiently watered low-lying clay soils and to the river banks. *Acacia albida* is closer to the main channel than *A.seyal* whether in riverain soils or in wadis. A hypothetical transect along a major wadi would show the following relationships between the habitats of five common *Acacia* species: *A.tortilis* can tolerate very austere conditions and is associated with high ground from which water runs off easily and hence has less available water than *A.nubica* and *A.ehrenbergiana; Acacia seyal* is generally further from the channel than *A. albida* and *A.nilotica,* for its water needs are less than the latter two species.

The grasses are mainly annual species; heavy grazing and the low rainfall do not allow the grasses to grow to the stage where fires can become an annual possibility. In years of low rainfall (or in years following years of low rainfall) because of overgrazing there can even be an almost complete failure of annual plant growth. *Aristida* species are usually dominant and characterize areas with shallow surface deposits that are moistened in the brief rainy season, and places where the ground is relatively high and the surface deposits are coarse sand. *Schoenefeldia gracilis* replaces *Aristida* species on clay soils. Other ephemeral grasses present include *Sporobolus cordofanus, Dactyloctenium aegyptium, Eragrostis cilianensis, Cenchrus biflorus* and *Tragus berteronianus.* Heavy grazing seems to have allowed only vestiges of the perennial sand-binding grass *Panicum turgidum* to persist on sandy drainage lines. However, in heavily overgrazed areas and on clay soils it is sometimes replaced by the unpalatable *Cymbopogon proximus,* especially in the north.

With the grasses there is a low proportion of herbs: *Indigofera, Boerhavia* and *Tephrosia* spp., *Tribulus terrestris, Ipomoea cordofana* and *Trianthema pentandra* form the main annual herbaceous cover, whereas *Cassia senna, C. italica, Indigofera oblongifolia* and *Aerva javanica* are the main perennials. The two *Cassia* species occur where the vegetation is disturbed through extensive overgrazing or by the subsequent erosion of the soil. These species grow abundantly in areas where the soil is disturbed by digging for mud-brick making or for other purposes. Being unpalatable and tolerant of high temperatures and drought, these *Cassia* species flourish throughout the year, replacing the former vegetation.

8.2 SEMI-DESERT GRASSLAND ON CLAY

Harrison & Jackson (1958) considered that before the development of irrigation part of the Central Gezira was semi-desert grassland. The mean annual

---

b) Riverain scrub and woodland:

3. Riverain woodland with dominant *Acacia tortilis.*
8. Riverain woodland with *Capparis decidua, Ziziphus spinachristi, Acacia tortilis* subsp. *raddiana, A.seyal* and *A.nubica* dominating in various associations.
9. Riverain woodland in which *Acacia nilotica, A.seyal, Balanites aegyptiaca* and *Ziziphus spina-christi* are associated in various degrees of dominance.

rainfall increases southwards from about 250 mm to about 360 mm at Wad Medani. The bulk of this rainfall falls during July, August and September with the peak in August (Figure 1). Temperatures are generally slightly lower than those of the semi-desert scrub with a mean minimum temperature of 14 °C and a mean maximum of 33 °C for January, and a mean minimum of 24 °C and a mean maximum of 41 °C for May at Wad Medani.

Apart from the areas near the Nile banks, where erosion has led to the formation of hard-surfaced gravelly and sandy soils, the soil of the Gezira is a heavy alkaline cracking clay, with a clay fraction of about 60 % and a pH of about 9. On drying, the soil shrinks considerably and a network of wide and deep cracks is formed. After rainfall the upper layers of the soil absorb moisture, swell, and become almost impermeable. Sustained rainfall leads to flooding even on level ground. However, a break of two to three days is sufficient to cause the soil to dry, crack and become hard. The only water courses are a few small khors which originate from the high areas of hard-setting clays with no or rare fine cracks.

Along the Nile banks, where the soil is eroded, the woody vegetation is confined to the flood-plain and to the watercourses and catchment areas. The lower part of the flood-plain has *Acacia nilotica* and *A.seyal,* with *Ziziphus spina-christi* and *Balanites aegyptiaca* on the higher ground. On relatively large catchment areas adjacent to the higher ground *Acacia tortilis* subsp. *raddiana* is dominant; isolated occurrences of this tree are frequently found across the Gezira plain. *Capparis decidua* is dominant further away from the banks, especially along the beds of khors and wadis. Indeed, Mahmoud & Obeid (1971) reported that *Capparis decidua* is found both in the sufficiently watered low-lying clay areas and in the shallow wadis draining these sites. In most of these localities *Acacia nubica* is to be found either as pure stands or associated with *Capparis decidua.* Mahmoud & Obeid (1971) noted that *Acacia nubica* was found in the upstream parts of the principal wadis or on lower ground which receives runoff water from adjacent areas.

Around villages, where the grazing pressure is high, *Calotropis procera* is dominant together with *Cassia senna, C.italica* and *Acacia nubica.* There is a certain similarity between *Calotropis* and *Cassia* community types: both seem to flourish following destruction of the original natural vegetation. Both are abundant on disturbed soils, but *Calotropis procera* is not found in gravelly habitats and is almost always associated with alluvial sands and loams. The soils found in the habitats of *Cassia senna* may be homogeneous like those of the grassland types or may be of mixed material (sand and gravel). *Cassia senna* community type seems to have a wide range of moisture tolerance including rill channels on erosion pavements, shallow desert wadis, the sides of drainage ditches and river banks. *Calotropis procera* seems to be favoured in habitats wetter than those of *Cassia;* and is mostly confined to the river banks and fallow fields of irrigated land. When found away from river banks, it is restricted to sufficiently watered low-lying areas or to drainage ditches. The wide distribution of this species seems to be attributable partly to its parachute-equipped seeds that are easily dispersed by wind, and partly to its unpalatability.

On the heavy alkaline cracking clay soils the natural vegetation in the Central Gezira is grassland without trees or shrubs. Areas with *Acacia mellifera,* though considerable in total, are confined to narrow units along khors. Under higher rainfall, *Acacia mellifera* is dominant on these cracking clays and there is a gradual transition to *A.mellifera* thornland at about the 400 mm isohyet. The only other bush occurring in quantity is *Acacia nubica* on poorer sites, on lower ground, and as an indicator of overgrazing.

Ground cover comprises mostly grasses and includes *Aristida* spp., *Schoenefeldia gracilis, Cymbopogon nervatus, C.proximus, Sehima ischaemoides, Tragus berteronianus, Dactyloctenium aegyptium, Eragrostis cilianensis* and *Sporobolus cordofanus.* Other herbs include *Ipomoea* spp., *Cleome viscosa, Indigofera* spp., *Blepharis edulis* and *Trianthema pentandra.*

8.3 ACACIA THORNLAND ALTERNATING WITH GRASS AREAS

This area lies south of the line joining Tagra and Wad Medani and extends southwards to the line joining Kosti, Jebel Moya and Sennar. The annual mean rainfall ranges from about 280 mm at Ed Dueim to about 400 mm at Sennar. Although the bulk of this rainfall falls between July and September, with a peak in August (Figure 1), May, June and October contribute about 25 % of the total annual rainfall. Temperatures at Sennar are generally similar to those of the central belt described above with a mean minimum temperature of 15 °C and a mean maximum of 34 °C for January, and a mean minimum of 24 °C and a mean maximum of 41 °C for May.

The greater part of the area is covered by dark alkaline cracking clays that were laid down under permanent water (Tothill 1948). Other soil types occupy a fraction of the area. These include the localized grey skeletal and red and brown soils which form part of the catena developed on and around the scattered rock outcrops and jebels; the sandy loams and clays of the eroded slopes next to the Blue and White Nile channels; and the more recently deposited alluvial *gerf* sands, sandy loams and clays of the Blue and White Nile valleys.

The vegetation of this area can be subdivided into riverain associations; and *Acacia* Thornland on dark cracking clay alternating with grass areas.

8.3.1 *Riverain associations*

The vegetation along the banks is a mixture of *Acacia nilotica* on the flood plain with *A.seyal, Ziziphus spina-christi* and *Balanites aegyptiaca* in various degrees of abundance. In the northern part *Capparis decidua* occurs in pure stands alternating with *Acacia nubica* which occurs specifically around villages and in low-lying areas. *Acacia ehrenbergiana* and *A.tortilis* subsp. *raddiana* are of rare occurrence. The date palm *Phoenix dactylifera,* which is an indication of settlements, is found near villages or former settlement sites. *Mimosa pigra* is associated with the river bank and is always very close to the water.

Along the lagoons and banks of the White Nile *Eichhornia crassipes, Cyperus rotundus, Cyperus* spp., *Echinochloa* spp. and *Typha domingensis* are common. On the higher ground of the banks, *Chrozophora plicata, Leonotis africana, Tribulus terrestris, Heliotropium* spp., *Trianthema pentandra* and *Aerva javanica* occur. On watercourses which drain the higher ground and where sandy deposits prevail, tussocks of *Panicum turgidum* are common. The ground vegetation along the Blue Nile bank comprises most of the species mentioned above except *Eichhornia crassipes,* in addition to *Borreria phyteuma, Commelina kotschyi, Achyranthes aspera* and *Cleome viscosa,* which appear after the first rains and are to be found in all depressions and wadis between Kosti and Jebel Moya.

8.3.2 *Acacia Thornland on dark cracking clay alternating with grass areas*

This community occurs on clay soils under a rainfall range of between 400 and 570 mm approximately. Below about 400 mm there is a gradual transition to Semi-Desert Grassland on clay, and above about 570 mm there is a transition to *Acacia seyal-Balanites* Savanna (Harrison & Jackson 1958).

The vegetation in the north consists of *Acacia mellifera* and *A.nubica* and *Capparis decidua* thickets, and in the south of *A.mellifera, Boscia senegalensis* and *Cadaba rotundifolia* with scattered *A.seyal, A.senegal, Balanites aegyptiaca* and open areas of grasses and herbs. *Acacia mellifera* is confined to depressions and catchment areas in almost all sites. It is continuously being exploited to provide new lands for cultivation. Because of its gum arabic production *Acacia senegal* is not destroyed. In contrast to *Acacia senegal, A.seyal* is regularly cut for charcoal making. Individuals of this species are saved till they are six to eight years old and then cut. *Balanites aegyptiaca* is widespread in the area, because it is fire-resistant, hard to cut, does not produce a good quality charcoal and has edible fruits which are collected and sold.

Associated with *Acacia mellifera* are *Cadaba glandulosa, C.rotundifolia* and *Boscia senegalensis* which often persist after the *A.mellifera* has disappeared.

The most abundant species in the grassy areas are *Cymbopogon nervatus, Sorghum purpureo-sericeum, Hyparrhenia rufa, Sehima ischaemoides, Tetropogon cenchriformis, Setaria verticillata, Dinebra retroflexa, Dactyloctenium aegyptium,* and *Cenchrus ciliaris.* The herbs include *Cassia* spp., *Solanum dubium, Ipomoea cordofana, Rhynchosia minima* var. *memnonia,* and *Achyranthes aspera.*

We will now discuss in further detail the three main azonal associations to be found in this region.

8.4 THE SWAMP VEGETATION

The aquatic and riverine vegetation of a major river such as the Nile is not greatly affected by local climatic changes, provided the supply of water is not affected. We distinguish areas that are subject to flooding by the Nile from peripheral areas that, for topographical reason, rely on runoff for their water supply. The latter areas are obviously subject to local climatic changes. Swamp conditions exist whenever the land is permanently under water.

During the past two decades the introduction and population explosion of the water hyacinth, *Eichhornia crassipes,* a free floating aquatic from South America, has severely impeded the flow of the middle reaches of the Nile to the south of Khartoum (Gay & Berry 1959). The effect of this weed on the flora of the Sudd has not been recorded. Since the purpose of this chapter is to describe the existing vegetation and its reaction to the present environment in order to throw some light on former changes in vegetation, the effect of such an obvious alien would tend to confuse the issue. The following description is therefore of the Sudd prior to the first recording of the water hyacinth in March 1958. Some consideration is also given to establishing the status (native or otherwise) of the constituent species.

The vegetation of the Sudd has been described by Chipp (1930), Andrews (1948), Migahid (1948), the Jonglei Investigation Team (1954) and Rzóska (1974). The dominant species is the papyrus, *Cyperus papyrus,* which grows in dense clumps along the muddy banks of the rivers and lagoons as well as in the stagnant waters of the deep swamps. The species is not recorded north of about Kosti (Andrews 1948), although papyrus rafts have been seen floating past Khartoum. It is not recorded along the Blue Nile.

Tackholm & Drar (1950) believe that papyrus was once widespread in Egypt during Pharaonic times. They regard its almost total disappearance from Egypt as being largely due to siltation, drainage, and intensive cultivation along the banks of the Nile. Hadidi (1971) has reported what is regarded as a relict habitat for papyrus in the Wadi Natron depression. Thus the available evidence suggests that this widespread species is indeed native in the Sudan.

The papyrus is locally interspersed by the grass *Vossia cuspidata,* which extends into the water for a considerable distance by means of its submerged rhizomes. It usually occurs in pure stands fringing the main river and other khors. *Vossia* is also present in the Blue Nile, and is a troublesome weed of the Gezira canals but apparently does not occur to the north of Khartoum. Widely distributed through the Old World tropics, it is probably native in the Sudan.

Another widespread tropical grass that is of importance in the Sudd is the tall, reed-like *Echinochloa pyramidalis,* either associated with *Vossia* or forming locally pure stands. North of the Sudd it has been recorded from Ed Dueim.

The reed grasses, *Phragmites karka, P.mauritianus* and possibly *P.communis* dominate large areas of seasonally inundated parts of the Sudd. Taxonomically, *Phragmites* is a difficult genus. The species are not easy to separate and have not been sufficiently collected. In the Sudan the situation is further complicated by introgression between the two tropical species and the temperate *Phragmites communis.* Both *Phragmites karka* and *P.mauritianus* have been recorded as far north as the Egyptian border (Ahti et al. 1973), while *P.communis* has been known in Egypt since the First Dynasty (Tackholm & Drar 1950).

The reed mace or bulrush, *Typha domingensis,* often forms small isolated stands in habitats rather similar to those of *Phragmites.* Its distribution extends northwards into Egypt. The Neolithic record from Ancient Egypt (Tackholm & Drar 1941) may perhaps refer to this species.

Ambatch, *Aeschynomene elaphroxylon,* is widely distributed in the Sudd, usually occurring in clumps in areas subject to seasonal inundation. The local inhabitants construct canoe-rafts from bundles of the balsa-like stems. Its northernmost recorded locality is Abba Island (Broun & Massey 1929). This too may be regarded as native.

Water lettuce, *Pistia stratiotes,* is a pantropical free-floating aquatic herb. In the Sudd it often forms dense mats across the surface of shady, still or slow-flowing waters; it is also known from the Gezira canals. It is not known for certain whether the plant existed in Egypt during the Pharaonic period (Tackholm & Drar 1950); its present rather rare occurrences in Egypt are probably due to recent, sporadic introductions. Its early presence in the Sudd should therefore be regarded as uncertain.

Other associated aquatic herbs are *Ipomoea aquatica, Nymphaea lotus, Ludwigia stolonifera, Ottelia ulvifolia, Potamogeton* spp., etc., a number of which also occur in the Gezira canals. All these plants may well have existed in the ancient swamps of the White and Blue Niles, and some which are now found along the canals of the Gezira today are possibly re-introductions to the area.

The Sudd is also the home for such swamp-loving mammals as the Nile lechwe, situtunga and the hippopotamus. The Nile lechwe is entirely restricted to the Sudd, an important fact when considering the evidence of the fossil fauna as a means of interpreting past environments, as noted in chapter 3. Other mammals that visit the Sudd for varying periods include the elephant, buffalo, tiang, kob and reedbuck (Cloudsley-Thompson 1966; Rzóska 1974).

The seasonally inundated areas surrounding the Sudd are known as 'toich'. Although the annual flooding is sufficient to inhibit tree growth, it is not so long as to encourage swamp conditions. These water-meadows are dominated by pure stands of such grasses as *Vetiveria nigritana, Sporobolus pyramidalis, Oryza barthii,* etc. Where the ground is sufficiently elevated so that it is rarely flooded, although the water-table is high, the palms, *Hyphaene thebaica* and *Borassus aethiopum* may be locally abundant. Elsewhere *Acacia seyal* and *Balanites aegyptiaca* or even *Acacia sieberana* may be the dominant trees. It too is an area that is rich in game.

8.5 RIVERINE WOODLAND

Pure stands of *sun't, Acacia nilotica* subsp. *nilotica* and subsp. *tomentosa,* are a familiar sight along the banks of the Nile, from the Sudd northwards into Egypt. These *sun't* forests are far less extensive today due to widespread felling for boat building, fuel, and railway sleepers (Muriel 1901); the trees were even used to pay taxes (Schweinfurth 1893). *Sun't* thrives in areas that are seasonally inundated, but cannot tolerate permanent flooding or dry-land conditions. Drar (1970) has suggested that subsp. *nilotica* is unable to withstand such long periods of flooding as subsp. *tomentosa.* A characteristic locality for these trees is the silt-filled ox-bow lake depressions along the Blue Nile. The colonization of these depressions has been described by Chipp (1930).

The first trees to colonize a new-formed sandbank in the river are *Salix subserrata* and *Tamarix nilotica.* As the banks are built up by further deposition of sand, more xerophytic species such as *Ziziphus spina-christi* become established (Figure 3). Presumably the ox-bow lakes behind the sandbanks become reed swamps and, as they silt up and become dry, *Acacia nilotica* will

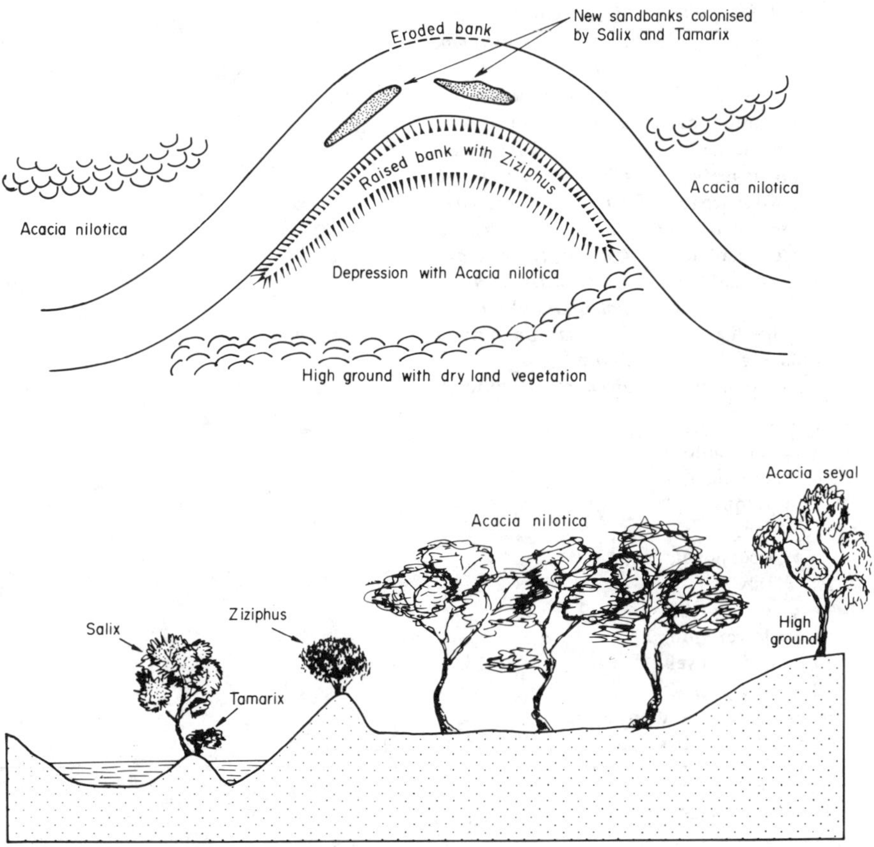

Figure 8.3. Transect across the Blue Nile showing the relationship between topography and vegetation (adapted from Chipp 1930).

Table 8.1. A list of the plant species and the state of their abundance at Jebel et Tomat and Jebel Moya.

		Jebel et Tomat			Jebel Moya
		Eastern jebel	Western jebel	Col between the two jebels	
Capparidaceae	*Cadaba rotundifolia*	w	d***		
	Cleome viscosa	w	w		w
	Maerua sp.		r		
Amaranthaceae	*Aerva javanica*		vr		
Combretaceae	*Combretum aculeatum*	r	r		
	Combretum hartmannianum				ld
Tiliaceae	*Grewia tenax*	c	d**, w		c
Sterculiaceae	*Sterculia setigera*				f
Malvaceae		r			
Leg.-Mimosoideae	*Acacia albida*				f
	Acacia ehrenbergiana				f[4]
	Acacia mellifera	c	d*	w	
	Acacia nubica	f	r	w	f[4]
Leg.-Papilionoideae	*Tephrosia* sp.	r	o		
Rhamnaceae	*Ziziphus spina-christi*				c[3]
Ampelidaceae	*Cissus quadrangularis*				o
Simaroubaceae	*Balanites aegyptiaca*			w	f[4]
Burseraceae	*Boswellia papyrifera*				f
Asclepiadaceae	*Calotropis procera*	l[1]		lw	f[4]
Solanaceae	*Lycium persicum*	f	f		o
	Solanum dubium	r			
Pedaliaceae	*Rogeria adenophylla*	w			w
Liliaceae	*Dipcadi* sp.				c
Gramineae	*Beckeropsis uniseta*	f[2]	f		

Key:
c = common
d = dominant
f = frequent
l = local
o = occasional
r = rare
w = widespread
1 on lower ground
2 at higher levels
3 on the plain
4 in valleys and depressions
d*, d**, d*** first, second and third order of dominance

then be able to establish itself. Eventually the depressions are completely filled by silt, the *Acacia nilotica* dies and is replaced by more mesophytic species, typical of the surrounding dry-land area. This natural succession has been observed by one of us (G.E.W.) in both Kordofan and Darfur.

8.6 VEGETATION ON JEBELS

Table 1 lists the plant species recorded at Jebel Tomat and Jebel Moya. The distribution of vegetation on the jebels seems to be mainly controlled by precipitation, slope, and the degree of rock weathering. The germination, establishment and growth of many plant species on the different sites appear to be a function of the accumulation of fine-textured material. Seepage increases or decreases the amount of runoff received by many sites.

The presence of *Combretum hartmannianum* and *Sterculia setigera* is of interest here. The former represents a northern outlier from its main area of occurrence, suggesting possible isolation arising from climatic change. The latter is probably at or very near the northern limits of its distribution range in terms of soil moisture tolerance.

The foot of the steep slopes and the adjacent plain form large catchment areas with a water budget very close to that of the riverain associations. The plain at the foot of the steep slopes of Jebel Moya has *Ziziphus spina-christi, Balanites aegyptiaca, Acacia nubica, Calotropis procera, Setaria incrassata, Hyperrhenia rufa, Vetiveria nigritana, Cissus quadrangularis, Boerhavia* and *Cyperus* spp.

8.7 DISCUSSION

The vegetation of the Gezira is unexceptional in that it has a past history, a present status, and a future development. The area has a long history of association with man. The activities of cultivators from Neolithic times onwards have caused considerable destruction of the original vegetation, mainly through the practice of 'fire-stick farming'.

Fire, the axe and the hoe were employed to clear the land, which was then cultivated for a few seasons. Cultivation of the soil has often eliminated much of the original vegetation and has enabled secondary plant succession to proceed. In the Gezira, this secondary succession is represented by a number of tree species, chiefly certain Acacia species, with *Capparis decidua, Ziziphus spina-christi* and *Balanites aegyptiaca* as faithful allies.

A certain degree of intuitive reasoning combined with a wide knowledge of similar vegetation types elsewhere is therefore necessary in order to reconstruct the palaeoflora, as attempted in chapter 3.

The physical environment seems to control the nature of the ecosystem in the riverain associations dominated by *Ziziphys spina-christi, Acacia nilotica* and *A.seyal;* in the overgrazed areas dominated by *Acacia nubica* and

Calotropis procera; and in the depressions and catchment areas colonized by *Acacia mellifera.* On rocky hillslopes rainfall, erosion, seepage and the accumulation of fine-textured material seem to have provided a variety of micro-habitats that have enabled a number of plant species to establish and grow, which accounts for the relatively rich flora of these jebels.

The vegetation in the southern part of the Gezira is believed to have been composed mainly of Combretaceous species (Kassas 1968). Fire and cultivation would seem to have changed the composition of the original vegetation. Kassas (1968) describes the situation as follows: 'The first fire destroys beyond regeneration: *Piliostigma reticulata, Entada sudanica, Stereospermum kunthianum, Acacia sieberana* and *Dichrostachys cinerea* (syn. *D.glomerata*). The half-burnt trees of *Combretum hartmannianum* seem capable of coppice regeneration and may manage to keep their stand. Repeated fire will destroy *Combretum* and the original forest passes to a phase of open growth of *Balanites aegyptiaca* and *Acacia seyal*.'

Also, there seems no good reason for believing that the present southward shift of the major vegetational belts in the Sudan, to which Kassas (1968) has drawn attention in a comparison between the present ecological boundaries and those shown by Smith (1944) and by Harrison & Jackson (1958), is due entirely to changes in rainfall. The evidence (Obeid & Seif El Din 1970, Chialvo 1975) supports the view that the increased pressure in land use for cultivation and grazing is a major factor in the deterioration of the vegetation.

We have emphasized the destructive influence of agricultural and pastoral man upon the original vegetation of the central Sudan. The economic importance of the plant cover far transcends its value as a source of fuel or fodder, for many plants are widely used for food, medicine and other domestic purposes. The list of such plants and their uses given in Table 2 is by no means exhaustive, and shows that far more extensive use is made of the vegetation by the rural inhabitants of the central Sudan than is generally realized. We hope that such information will be of value to the palaeoethnobotanist, and so have included a number of species not recorded in this area today, but which do occur further south, and so were probably present during the early Holocene.

In this chapter we have noted the influence of soil type, drainage and micro-relief, all of which are of course interrelated, upon the present-day distribution of plant communities and species in central Sudan. Superimposed upon these physical factors is the overwhelming influence exerted by man. Through burning, clearing, cultivation and grazing man has so changed the plant cover during the past few thousand years that very little of the original vegetation now exists in the Gezira, except possibly for a few species on isolated rocky hills in the far south. As the most densely-settled agricultural area in the Sudan, the Gezira is inevitably the least promising locality from which to attempt a reconstruction of the early Holocene flora. However, sufficient tracts of comparatively undisturbed vegetation still persist to the west, south and east of the Gezira to have encouraged us to attempt the exercise described in chapter 3.

Table 8.2. List of economically useful plants in modern Sudan.

Economic plants: a large number of plants are of local economic importance. A selection of species from Bacon (1948), Hunting Technical Services (1964a, b, 1968), Boulos (1966), Baumer (1975) and Chialvo (1975) in the area under consideration are listed below, together with their use by man. The list is by no means exhaustive.

Plant family	Plant species	Economic use
Nymphaeaceae	*Nymphaea lotus*	seeds and roots eaten
Capparidaceae	*Boscia salicifolia*	fruit and roots edible
	Boscia senegalensis	fruits sometimes eaten, leaves unpalatable to stock
	Cadaba rotundifolia	used as a purgative
	Capparis decidua	fruit edible; used for dressing boils and swellings
	Cleome monophylla	leaves eaten in salads
	Cleome viscosa	leaves eaten in salads
	Crataeva adansonii	leaves and fruit eaten
	Gynandropsis gynandra	leaves eaten in salads
Portulacaceae	*Portulaca oleracea*	leaves eaten in salads
Amaranthaceae	*Aerva javanica*	roots used for toothbrushes
	Amaranthus spp.	leaves used as a vegetable
	Celsia argentea	seeds used for treating dysentery
	Celsia trigyna	leaves used as a vegetable
Tamaricaceae	*Tamarix nilotica*	wood used for fuel
	Tamarix orientalis	galls used for tanning and dying
Cucurbitaceae	*Colocynthis citrulus*	seeds lightly roasted and chewed; fruit a source of water for humans and stock during the dry season
	Colocynthis vulgaris	tarry liquid distilled from seeds used for proofing waterskins
	Luffa cylindrica	fruit fibres used for cleaning squabs
Combretaceae	*Anogeissus leiocarpus*	tea-like beverage made from leaves; useful timber, wood ash used in dying
Tiliaceae	*Corchorus olitorius*	leaves used in soups
	Grewia flavescens	fruit edible
	Grewia mollis	fruit edible
	Grewia tenax	fruit edible, leaves used as a vegetable
	Grewia villosa	fruit edible; wood used for spear shafts and arrows
Bombacaceae	*Adansonia digitata*	fruit edible, young leaves used as vegetable
Sterculiaceae	*Sterculia setigera*	young leaves and gum edible
Malvaceae	*Hibiscus sabdariffa*	drink made from the accrescent calyx; stem fibres used for cordage
	Sida alba	used in the treatment of snake bites

Table 8.2. (continued)

Plant family	Plant species	Economic use
Euphorbiaceae	*Euphorbia candelabrum*	latex used as an arrow poison
	Euphorbia polycnemoides	used to stimulate lactation
	Euphorbia venefica	latex used as an arrow poison
	Euphorbia aegyptiaca	decoction used for headaches and fractures
	Ricinus communis	leaf compress used for headaches
	Securinega virosa	bark used for fish poison
Leg.-Caesalp.	*Bauhinia rufescens*	bark used as cordage
	Cassia italica	seeds used as vegetable and medicine
	Cassia occidentalis	boiled leaves used as a febrifuge
	Cassia senna	fruit a laxative, used for treating dysentery
	Piliostigma thonningii	seeds used for soup; bark used for cordage
	Tamarindus indica	fruit edible
Leg.-Mimos.	*Acacia mellifera*	wood used for throwing sticks; roots used for basket making
	Acacia nilotica	seeds used for tanning, beverage and medicine; low grade gum
	Acacia nubica	roots used for cordage; bark used for tanning
	Acacia senegal	gum edible, used for making ink
	Acacia seyal	roots used for baskets; gum edible
	Acacia sieberana	decoction of the fruit used medicinally
	Acacia tortilis subsp. *raddiana*	trunk used for butcher's blocks
	Albizia anthelminthica	paste from bark used as vermifuge
	Dichrostachys cinerea	seeds used in soups
Leg.-Papill.	*Aeschynomene elaphroxylon*	bundles of stems used for rafts
	Dalbergia melanoxylon	wood used for spear shafts and carving
	Tephrosia obcordata	a cooking oil extracted from the seeds
Ulmaceae	*Celtis integrifolia*	fruit edible
Moraceae	*Ficus iteophylla*	fruit edible
	Ficus sycomorus	fruit edible; bark used for kitchen utensils
	Ficus spp.	latex used for bird lime
Salvadoraceae	*Salvadora persica*	twigs and roots used for toothbrushes; fruit edible, seeds give an illuminating oil
Olacaceae	*Ximenia americana*	fruit edible
Rhamnaceae	*Ziziphus mauritiana*	fruit edible
	Ziziphus mucronata	fruit edible but bitter; wood used for throwing sticks
	Ziziphus spina-christi	fruit edible

Table 8.2. (continued)

Plant family	Plant species	Economic use
Ampelidaceae	*Cissus quadrangularis*	boiled stem used for treating stomach ache
Simaroubaceae	*Balanites aegyptiaca*	fruit edible; cooking oil extracted from the crushed leaves; wood used for throwing sticks, pestles and mortars, platters, saddles and writing slates; boiled leaves eaten in times of famine
Burseraceae	*Commiphora africana*	root edible, gum fragrant
Anacardiaceae	*Sclerocarya birrea*	fruit laxative
Asclepiadaceae	*Calotropis procera*	latex used as arrow poison, leaves used in beer fermentation
	Leptadenia pyrotechnica	stems used as cordage
Rubiaceae	*Xeromphis nilotica*	fruit used as fish poison
Compositae	*Eclipla alba*	produces a black dye used for tattooing
	Lactuca capensis	leaves eaten in salads
	Sonchus spp.	leaves eaten in salads
Boroginaceae	*Cordia rothii*	fruit edible; fibre used for cordage
	Heliotropium indicum	used medicinally
Solanaceae	*Solanum incanum*	ripe crushed fruit used to curdle milk for making cheese, leaves used medicinally
Convolvulaceae	*Ipomoea aquatica*	young shoots edible
Acanthaceae	*Blepharis linariifolia*	favourite food of the Oryx
Labiatae	*Ocimum basilicum*	refreshing drink from an infusion of the seeds
Liliaceae	*Urginea indica*	bulb rubbed on the skin as a counter-irritant
Palmae	*Borassus aethiopum*	fruit edible; fresh sap a pleasant drink
	Hyphaene thebaica	fruit edible; pounded kernel used as flour
Cyperaceae	*Cyperus rotundus*	tubers edible
Gramineae	*Anthephora* spp.	seeds harvested for grain in lean years
	Aristida spp.	stems used as brushes
	Brachiaria obtusiflora	seeds harvested for grain in lean years
	Cymbopogon spp.	thatching grass
	Dactyloctenium aegyptium	seeds harvested for grain in lean years
	Digitaria spp.	seeds harvested for grain in lean years
	Echinochloa pyramidalis	seeds harvested for grain in lean years
	Echinochloa spp.	seeds harvested for grain in lean years

Table 8.2. (continued)

Plant family	Plant species	Economic use
	Eleusine indica	seeds harvested for grain in lean years
	Eragrostis tremula	stems woven into mats
	Eragrostis spp.	stems used as brushes
	Hyparrhenia	thatching grass
	Oxytenanthera abyssinica	seeds harvested for grain in lean years; stems used for spear handles
	Panicum turgidum	seeds harvested for grain in lean years
	Panicum spp.	seeds harvested for grain in lean years
	Pennisetum pedicellatum	thatching grass
	Phragmites spp.	stems used for building houses
	Rottboellia exaltata	seeds harvested for grain in lean years
	Schoenefeldia gracilis	stems used for brushes
	Setaria pallide-fusca	seeds harvested for grain
	Sorghum spp.	seeds harvested for grain in lean years; stems used for building houses
	Vossia cuspidata	stems used for thatching and mattresses

8.8 REFERENCES

Ahti, T., L. Hamet-Ahti & B. Pettersson 1973, Flora of the inundated Wadi Halfa reach of the Nile, Sudanese Nubia, with notes on adjacent areas. *Ann. Bot. Fennici* 10:131-162.

Andrews, F.W. 1948, The vegetation of the Sudan. In J.D. Tothill (ed.), *Agriculture in the Sudan.* Oxford University Press:32-61.

Bacon, G.H. 1948, Crops of the Sudan. In J.D. Tothill (ed.), *Agriculture in the Sudan.* Oxford University Press:302-400.

Baumer, M. 1975, Catalogue des plantes utiles du Kordofan (Rép. du Soudan), particulièrement du point de vue pastoral. *J. Agric. trop. et Bot. appl.* 22:81-119.

Boulos, L. 1966, Flora of the Nile region in Egyptian Nubia. Feddes Report 73:184-215.

Broun, A.F. & R.E. Massey 1929, *Flora of the Sudan.* Sudan Govt, Khartoum.

Bunting, A.H. & J.D. Lea 1962, The soils and vegetation of the Fung, east central Sudan. *J. Ecol.* 50:529-558.

Chialvo, N. 1975, Contribution à l'étude écologique de la végétation du confluent Atbara-Setit (Répub. Dém. du Soudan). Doc. 3e Cycle thesis, Grenoble.

Chipp, T.F. 1930, Forests and plants of the Anglo-Egyptian Sudan. *Geogr. J.* 75:123-143.

Cloudsley-Thompson, J.L. 1966, Climate and fauna of the central and southern Sudan. *Sudan Notes and Records* 47:127-136.

Drar, M. 1970, A botanic expedition to the Sudan in 1938. In V. Täckholm (ed.), Cairo Univ. Herb. Pub. 3.

El-Tom, M.A. 1972, The nature of rainfall over the Sudan and the potentialities for its artificial modification. African Seminar paper 15, Sudan Res. Unit, Univ. Khartoum: 45.

Gaitskell, A. 1959, *Gezira: a story of development in the Sudan.* London, Faber & Faber.

Gay, P.A. & L. Berry 1959, The water hyacinth; a new problem on the Nile. *Geogr. J.* 125:89-91.

Hadidi, M.N. el 1971, Distribution of *Cyperus papyrus* L. and *Nymphaea lotus* L. in inland waters of Egypt. *Mitt. Bot. Staatssamml. München* 10:470-475.

Halwagy, R. 1962a, The incidence of the biotic factor in northern Sudan. *Oikos* 13(1): 97-117.

Halwagy, R. 1962b, The impact of man on semi-desert vegetation in the Sudan. *J. Ecol.* 50:263-273.

Harrison, M.N. & J.K. Jackson 1958, Ecological classification of the vegetation of the Sudan. Republ. of Sudan, Min. Agric., Forests Dept. Bull. 2 (new series).

Hunting Technical Services 1964a, Land and water use survey in Kordofan Province of the Republic of the Sudan. Report on the geology, geomorphology and soils, vegetation and present land use. Doxiades Assoc. Athens Doc. Dox-Sud-A26.

Hunting Technical Services 1964b, Report 6, Roseires soil survey project, The White Nile, East Bank, Rabak to Khartoum. Soils and engineering reconnaissance.

Hunting Technical Services 1968, Land and water resources of the Jebel Marra area, Republic of the Sudan. Reconnaissance vegetation survey. LA:SF/SUD/17, FAO, Rome.

Jonglei Investigation Team 1954, The Equatorial Nile Project and its effects in the Anglo-Egyptian Sudan. Khartoum, Government Printer.

Kassas, M. 1956, Landforms and plant cover in the Omdurman desert, Sudan. *Bull. Soc. Géogr. d'Egypte* 29:43-58.

Kassas, M. 1968, Dynamics of desert vegetation. Proc. Tech. Meeting on Conservation and Ecology of the West Mediterranean, North Africa. Hammamet, Tunisia, March 1968 (mimeo).

Mahmoud, A. & M. Obeid 1971, Ecological studies in the vegetation of the Sudan. I. General features of the vegetation of Khartoum Province. *Vegetatio* 23(3-4):153-176.

Migahid, A.M. 1948, Report on a botanical excursion to the Sudd region. Fouad I. Univ. Cairo.

Muriel, C.E. 1901, Report on the Forests of the Sudan. Cairo.

Obeid, M. & Mahmoud, A. 1969, The vegetation of Khartoum Province. (Guide to the Natural History of Khartoum Province, Part IV). *Sudan Notes and Records* 50:134-159.

Obeid, M. & A. Mahmoud 1971, Ecological studies of the vegetation of the Sudan. II. The ecological relationships of the vegetation of Khartoum Province. *Vegetatio* 23 (3-4):177-198.

Obeid, M. & A. Seif El Din 1970, Ecological studies of the vegetation of the Sudan. I. *Acacia senegal* (L.) Willd. and its natural regeneration. *J. Appl. Ecol.* 7:507-518.

Oliver, J. 1965, The climate of Khartoum Province. (Guide to the Natural History of Khartoum Province, Part II). *Sudan Notes and Records* 46:90-129.

Omer, H.A. 1969, The ecology of the Fung Area. M.Sc. thesis, Univ. Khartoum.

Rzóska, J. 1974, The Upper Nile swamps, a tropical wetland study. *Freshwater Biol.* 4: 1-30.

Schweinfurth, G. 1873, *The heart of Africa. Travels and adventures from 1868 to 1871.* (transl. E.E. Frewer), 2 vols. London, Sampson Low & Co.

Smith, J. 1944, The grass acacia cycle in Gedaref and notes on fire protection. Report Soil Conservation Comm., Sudan Govt, Khartoum:87.

Smith, J. 1949, Distribution of tree species in the Sudan in relation to rainfall and soil texture. Sudan Govt Min. Agric. Bull. 4. Khartoum.

Täckholm, G., V. Täckholm & M. Drar 1941, Flora of Egypt. Fouad I. Univ. Bull. Fac. Sci. Pub. 1.

Täckholm, V. & M. Drar 1950, Flora of Egypt. II. Fouad I. Univ. Bull. Fac. Sci. Publ. 28.

Tothill, J.D. 1948, A note on the origins of the soils of the Sudan from the point of view of the man in the field. In J.D. Tothill (ed.), *Agriculture in the Sudan.* Oxford University Press: 129-143.

DONALD A.ADAMSON / MARTIN A.J.WILLIAMS
Macquarie University, North Ryde, Australia
RICHARD GILLESPIE
Sydney University, Australia

9

Palaeogeography of the Gezira and of the lower Blue and White Nile valleys

9.1 INTRODUCTION

The Gezira was formed by the Blue and White Nile rivers and its history is overwhelmingly that of the parent rivers. The Nile is not one river: it is a concatenation of rivers that spans more than 30 degrees of latitude. The Gezira, lying about halfway between its source and the sea, should provide evidence about palaeoenvironments and river history that is congruent with evidence throughout the whole drainage basin. The Gezira therefore occupies an important position for pan-African correlations. In a more restricted sense, a knowledge of local palaeoenvironmental conditions and of river history is necessary for interpreting prehistoric human occupation in central Sudan.

Although the alluvial fans of central Sudan probably contain a fairly complete Quaternary record, exposures of the sedimentary sequence are extremely restricted. In the eroded land ('kerrib') along the Blue Nile the upper 15 to 20 m have been exposed as a result of river incision. Exposures along the White Nile are even more restricted, particularly as a result of flooding of the White Nile valley upstream of the Jebel Aulia Dam, and on the Gezira plain itself erosional exposures do not occur. Our present ability to trace in detail the Quaternary history of the Gezira is limited to the late Pleistocene and Holocene as represented in the uppermost few metres of sediment. This situation can be contrasted with that along most of the Nile in Egyptian and Sudanese Nubia, where exposures are common (although many are now below the water of the Aswan Dam) but the sediments are fragmentary.

The dramatic difference in behaviour of the Blue and White Nile rivers has been detailed in chapter 7. The main Nile is best understood as a complex fusion of these two great streams. The same principle can be applied to the past. A study of the history of the separate Niles and of their sediments is a fruitful approach to understanding the history of the main Nile. We have followed this strategy in our study of the Blue and White Niles and of their associated landscapes south of Khartoum (Figure 1).

The pioneer studies of the palaeogeography and archaeology of the central Sudan carried out by J.D. Tothill (1946, 1948) and by A.J. Arkell (1949,

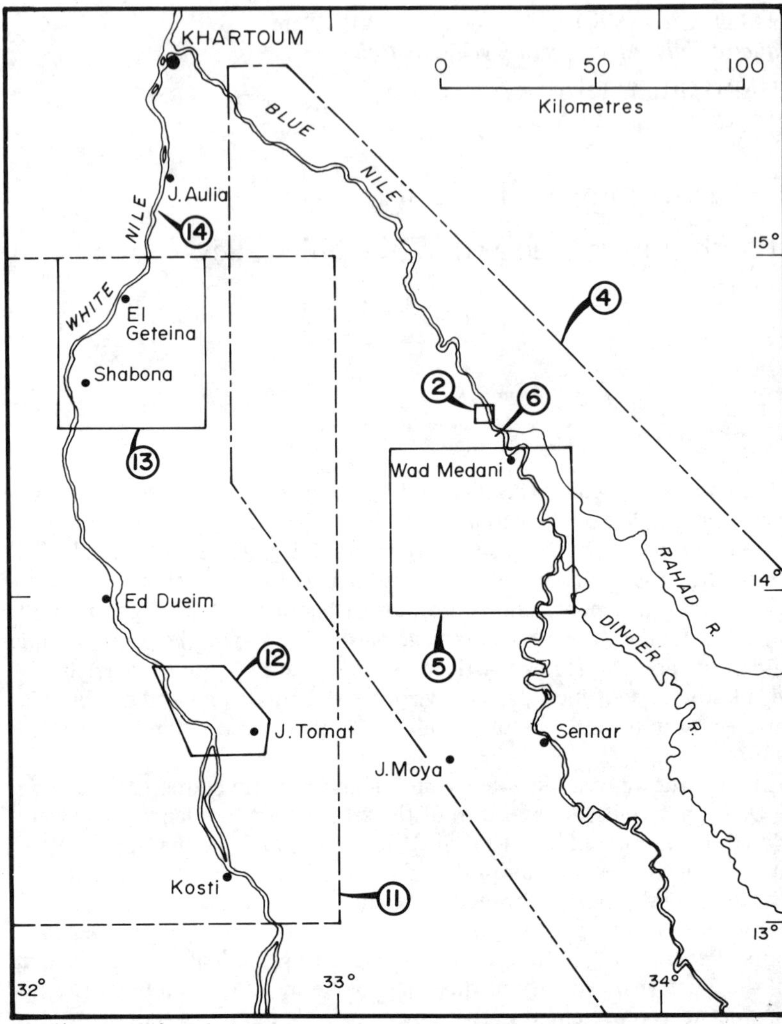

Figure 9.1. Location diagram for maps and diagrams in chapter 9.

1953) led to conclusions which have been vindicated by later work. Tothill's interpretation of the origin of the Gezira soils, based on his study of the distribution of sub-fossil mollusca, has been amply confirmed by the additional evidence that we have obtained. Tothill also reconstructed the palaeoenvironment of the Gezira with remarkable accuracy and subtlety. Arkell's work at Early Khartoum and Shaheinab was not only the foundation of modern archaeology in the Sudan: it also achieved two further remarkable results. Arkell was able to reconstruct accurately the palaeoenvironment around these sites and therefore on the Nile, and his observations led him to conclude that the Nile flood levels were about 5 m higher at the time of the settlement at Shaheinab c.5 000 years ago, and up to 10 m higher at the time of the Early Khartoum Culture at the Khartoum hospital site. Our work with Professor J.D. Clark now makes it clear that these latter flood levels were operative some 7 000 to 8 000 years ago (Adamson et al. 1974). Arkell's work therefore led us to the important inference that downcutting by the main Nile and its two major tributaries had occurred throughout much of the Holocene. Our subsequent work, spanning more than 10 years, has provided a framework of absolute dates and a substantial body of new evidence to support and extend this early work.

9.2 THE BLUE NILE IN THE LATE PLEISTOCENE AND HOLOCENE

Below the thin skin of Late Pleistocene and Holocene sediments, the Gezira fan consists of up to 110 m of sands and gravels with smaller amounts of clay (chapter 5). The chronostratigraphy of these deeper deposits is virtually unknown and their generalised age is probably broadly synchronous with the Cenozoic uplift and dissection of Ethiopia. Logging records for groundwater bores make it clear, however, that the Gezira contains sinuous sand and gravel beds many kilometres in length and up to 25 m thick (chapter 5, Figure 2) which undoubtedly correspond to the buried channels of former streams. Early distributaries of the former Blue Nile and Rahad and Dinder rivers commonly carried a fast and highly seasonal flow and a heavy load of local and Ethiopian sediment. Clay beds or lenses probably reflect episodes of overbank flooding similar to the Holocene interval during which the present surface clays were laid down. However, what must have been an extremely complex history of fluvial deposition, incision and lateral migration, repeated many times during the last few million years, cannot be interpreted with any degree of precision from the fragmentary bore-log records, particularly in the absence of any form of absolute dating.

The deeply buried sinuous channels are the ancient equivalents of the Late Pleistocene and Holocene prior stream channels exposed in the kerrib of the Blue Nile or visible on the present surface of the Gezira. These near-surface features can be studied in detail and dated fairly well and they supply much of the evidence with which we are concerned in this section.

9.2.1 *Volcanic ash deposits in the Gezira*

In January 1973, Adamson & Abdulla discovered deposits of remarkably pure volcanic ash, in the form of microscopic glassy fragments, exposed in the kerrib of the Blue Nile about 25 km northnorthwest of Wad Medani at kilo 163 beside the Khartoum-Wad Medani road. Similar deposits occur along the western side of the Blue Nile from just north of Wad Medani for at least 83 km to the village of Takala (Figures 2 and 3), and they are now being studied in more detail by Adamson, Abdulla & Frances Williams. The kerrib on the east bank of the Blue Nile has not yet been examined and areas south of Wad Medani have been examined incompletely without finding deposits.

The deposits occur in river channels which became choked with volcanic ash. Figure 2 shows one group of deposits about 12 km northnorthwest of

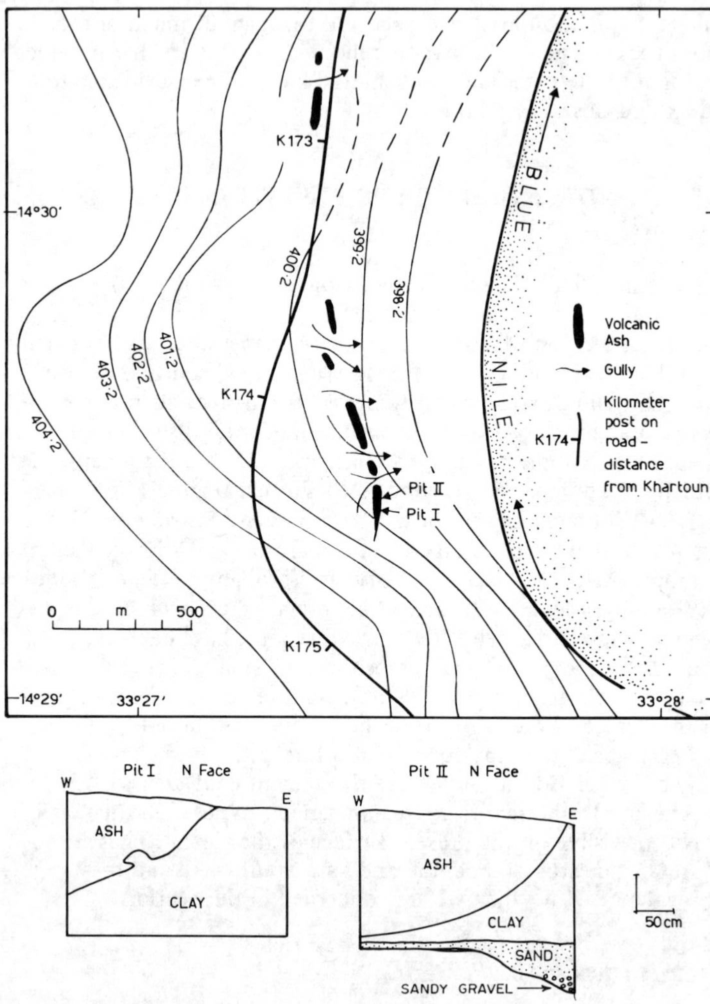

Figure 9.2. Volcanic ash deposits on the western side of the Blue Nile about 12 km north of Wad Medani. The deposits are exposed in the eroded land (kerrib) which lies between the surface of the Gezira plain at c. 404 m and the river. The northern faces of two pits dug across the ash-clay contact are also shown. The date of 24 920 ± 480 (SUA-605) was obtained on irregular carbonate concretions situated within the clay, 15 cm below the clay-ash contact in the western face of pit I.

Wad Medani between the villages of Wad el Magdoub and Fadasi Omarab. Numerous similar deposits are exposed elsewhere in the kerrib flanking the Blue Nile. Evidence for the fluvial origin of these deposits can be summarised: the deposits are sub-horizontal; deposits which occur close together are obviously part of a narrow linear feature which has been dissected by recent erosion on the slopes leading down to the entrenched Blue Nile; cross-sections show a sloping contact between the ash and the sediments in which the ash lies, typical of a stream channel which is not entrenched into its alluvial surface. These three lines of evidence are illustrated in Figure 2. Further evidence of the fluvial origin of these deposits is the finely laminated bedding which parallels the sloping contact between the ash deposit and the surrounding sediment, the fine cross-bedding, and the ripple marks which are revealed when the surfaces of laminae are exposed by erosion. The fine particle size may result partly from fluvial comminution of pumice. Rolled pumice up to 4 cm in diameter occurred in the near vicinity of the site at kilo 163 associated with impure ash. At Wad el Magdoub, about 10 km northnorthwest of Wad Medani, the ash is mixed with clay and biological remains such as phytoliths, sponge spicules and algal remains. Towards the northern (downslope) end of the Gezira, about 30 km west of Kamlin, H.H. Abdulla has shown that volcanic ash appears at the surface and at least in some instances forms the parent material of the Laota Sandy Clay Loam as defined by the UNDP Survey in 1968. It is interesting to note that this soil type often occurs in linear zones several kilometres in length.

The river channels in which the ash was deposited were shallow (3-4 m), relatively narrow (20-30 m) and relatively straight. The fluvial sedimentary features indicate rather gentle flow. In some cases the channels obviously contained abundant aquatic vegetation, particularly grasses which supplied the abundant phytoliths, and on which grew sponges and microscopic algae. These conditions are rather similar to those in modern irrigation canals and suggest slow-moving water occupying numerous distributaries. We suggest that massive ash and pumice eruptions in the Ethiopian headwaters blanketed part of the catchment and choked the rivers with a huge load of ash. Cut-and-fill structures are not obvious in the ash deposits, suggesting one continuous interval of ash deposition in the channels, under a low-energy flow regime. Continued deposition of the ash within the channels is also suggested by the preservation of irregularities in the river bank such as the slump feature shown in the northern face of Pit I (Figure 2) and by the absence of interbedding of ash and bank deposits or other sediments. It is, however, not known whether such channels were typical of the whole stream system at the time or whether we are seeing only overflow channels which were connected to a major channel located elsewhere. In any event, it is most unlikely that the fluvial regime was similar to that of the present Blue Nile in which a major river channel is incised into its fan and flow along minor distributaries cannot occur. It is more likely that flow was widely dispersed through numerous surface channels across the surface of the fan. The presence of the microfossils suggests that some at least of these channels contained permanent water.

The common occurrence of ash deposits downstream of the Rahad and Blue Nile confluence (Figure 3) suggests that the palaeo-Rahad may have supplied most of the ash which choked this part of the drainage system. If this conclusion is valid, it limits the search for a source to a relatively small part of the overall Blue Nile drainage basin. Another consequence would be that the palaeo-Rahad at this time followed a course, or courses, which were close to that of the present Blue Nile. The known ash channels are mostly sub-parallel to the present Blue Nile trough and the sites of the Laota Sandy Clay Loam soil are still well within reach of distributaries of a palaeo-Rahad. It must be recognised, however, that a thorough search has not yet been made for volcanic ash or ash-derived soils where the more westerly palaeo-channels of the Blue Nile system reach the distal parts of the Gezira fan in the general direction of El Geteina, Naima, and south of Jebel Aulia.

There is, of course, no reason to suggest that all the deposits of volcanic ash and other pyroclastics which occur in the Sudanese alluvial deposits are derived from the one series of eruptions. We believe that all the volcanic ash deposits we have so far mapped along the Blue Nile were deposited at about the same time, based on two radiocarbon dates to be discussed shortly. However, it is worth nothing that Ruxton (1956), quoted by Whiteman (1971), reported a 2 m thick layer of trachytic pumice at 13°19′N 34°25′E between the Rahad and Dinder rivers, at a depth of 14.4 m below the present surface. This is in contrast to the deposits in the Blue Nile kerrib which are only about 3 to 6 m below the present surface of the Gezira plain. It is, of course, possible but unproven that the Ruxton pumice deposit, the Blue Nile ash deposits and the Laota soils are all derived from one series of closely related eruptions and that the difference in depth of burial reflects a down-fan thinning of the more recent sediments.

However, multiple sources of pyroclastic deposits are certainly indicated as a result of the further discovery by Abdulla & Adamson in 1973 of a deposit of pure glassy volcanic ash 1.6 m below the surface of Holocene White Nile clays south of Kawa (Figures 11 and 12). This ash almost certainly entered the White Nile from the Sobat River or from the seasonal streams which drain from Ethiopia north of the Sobat. The deposit is possibly younger than the Blue Nile deposits since it lies just beneath a skin of Holocene clay.

Two radiocarbon dates from the Blue Nile deposits at two separate localities agree on a minimum age for the ash deposits (Table 1). A date of 23 780 ± 600 BP (I-9712) was obtained from fairly soft carbonate concretions formed *in situ* in clay, 15 cm beneath the bed of the ash-filled channel illustrated in Figure 2. At a second site at kilo 163 on the Khartoum-Wad Medani road near the village of Gereigreib and about 11 km northnorthwest from the previous site, a second date was obtained on thin horizontal bands of carbonate deposited between the ash lamellae. The sample was taken 40 cm below the ground surface and 20 cm above the ash-clay contact. This sample was dated at 24 920 ± 480 BP (SUA-605).

The closeness of both dates suggests that carbonate deposition within and beneath the ash-filled channels occurred at the same time at both sites.

Deposition of the two carbonates obviously occurred after the filling of the two channels with ash and their consequent abandonment. We consider that the time gap between ash deposition and carbonate precipitation was probably not particularly long, and we attribute a tentative age of c. 25 000 BP to the ash-filled channels exposed in the erosional flanks of the Blue Nile. If the Laota ash were deposited at about the same time it leads to the important conclusion that some of the superficial deposits towards the downslope end of the Gezira fan may be Late Pleistocene in age, whereas we shall see later in this chapter that other surface deposits on the Gezira are certainly early to middle Holocene in age. In other words, the surface of the Gezira plain (and of the Blue Nile fan) may be time-transgressive, with some deposits, particularly towards the distal end of the fan, being at least 10 to 20 thousand years older than the most recent surface sediments.

9.2.2 *Carbonate deposits and carbonate dates*

The use of carbonate to give an indication of the age of deposition of the volcanic ash raises the problem of interpreting radiocarbon dates obtained from carbonate in the central Sudan. The longitudinal section of the Gezira plain, revealed in the eroded trench of the present-day Blue Nile and to a lesser extent along the White Nile, has exposed many sediments containing abundant calcium carbonate in a variety of forms. Our observations on these carbonates have been made incidentally to our other research in the area, and a detailed study of the stratigraphy, age, and mineralogy of the deposits would reveal much about the Quaternary history of central Sudan and of the Nile. It would be particularly useful to be able to relate periods of carbonate deposition to local and regional palaeoclimate and to the behaviour of the Nile.

It is clear from our preliminary observations and from the available radiocarbon dates (Table 1) that the carbonates are of very diverse ages and have very different characteristics. The problems of interpreting the radiocarbon ages of carbonates are severe, but we feel that certain conclusions can be drawn tentatively from the available data.

Two fundamental problems are encountered. Firstly, the problem of distinguishing between carbonate deposited *in situ* (autochthonous) and reworked carbonate nodules which have been imported to a site by fluvial action like other sedimentary particles (allochthonous): obviously, a date on the former may be useful in giving information about the minimum age of the deposit in which the carbonate occurs. A date on allochthonous carbonate will give a maximum age for that deposit, which may not be very informative. Secondly, however, even with carbonate deposited *in situ*, the problem remains whether the carbonate being precipitated is in equilibrium with the atmosphere of the time or whether it is contaminated by older or younger carbon. Contamination by older carbon will introduce substantial errors, the deposit appearing much older than is actually the case, but even 20 % contamination by inert carbon will still increase the apparent radiocarbon age by only 1 700 BP, which is not a serious error for samples with a 'true' age of 20 000 BP. Far more serious is the problem of contamination by modern carbon, 5 % of which will reduce a true sample age of 30 000 to 21 000 BP, and 20 % of which will lower it to 12 200 BP (Polach & Golson 1966). In an alluvial fan such as the Gezira, in an arid to semi-arid environment, carbonate deposition, transport, solution and redeposition have probably been active throughout the Quaternary. We have made a preliminary attempt to assess these problems by sampling and dating various types of carbonate, particularly where mollusca also occur in the same deposit. Because of the expense involved, and the more reliable dates that can apparently be obtained from mollusca, we have concentrated on dating shell and have not been able to investigate thoroughly the problem of carbonate dating in this region.

Carbonate nodules which have been transported from their site of formation and redeposited can often be distinguished from *in situ* carbonate. Reworked nodules often occur as hard, rounded or sub-rounded pieces ranging widely in size from very coarse sand (c. 1 to 2 mm) to coarse pebbles or even cobbles c. 10 cm in diameter. Thin-section examination of some of the older carbonate pebbles shows that they have been reworked several times. The lack of a genetic relationship between reworked carbonate and the surrounding sediment is usually very obvious. For example, carbonate gravel in a sandy matrix, the whole showing clear fluvial bedding, is clearly an example of redeposition. Examples are shown in Figures 6 and 14 and in the Singa and Abu Hugar carbonate gravels described by Arkell and by Berry (Whiteman 1971:123-124). Our own dates for I-9291 (>40 000 yr BP) and I-9292 (12 640 ± 180 yr BP) are clear examples (Table 1 and Figure 14). Both samples came from the same layer, near the top of at least 2.5 m of fluvially-bedded carbonate gravels in a matrix of fine sand, the whole being overlain by up to 1 m of fine sand. This deposit was a palaeochannel, now exposed as a cliff by wave action from the present Jebel Aulia Dam. The older carbonate sample was composed of hard, dense, carbonate gravel, rounded or sub-rounded, black internally, and dark grey on the outer surface. The younger date was obtained from blocky lumps which were soft to medium hard, porous, angular to sub-rounded in shape, and whitish throughout. The older gravel gave every appearance of its antiquity and of reworking whereas the younger material would largely fragment and be dispersed if reworked.

Another example of allochthonous carbonate is illustrated from a pit in the Gezira Research Station at Wad Medani, situated in typical Gezira clay soil (Figure 8b). The shell date, 11 975 ± 260 BP (SUA-214), compares with the date on small hard carbonate nodules from the same depth of 27 370 ± 800 BP. We consider that the fine carbonate pebbles (c. 2 to 5 mm in diameter) which are hard, dense, dark grey to black throughout, and common in the Gezira clay soils (Finck 1961), are allochthonous. The dated soil horizon also contained some gypsum and diffuse carbonate which we presume to be autochthonous. In this case the problem exists of explaining the joint presence of clay and coarse-grained ancient carbonate nodules and the absence of particles of intermediate size class. The possibility exists that the fine carbonate gravel was transported by river action from a site where it had formed *in situ*, probably in clay, and that bed-load sands were not available for deposition at the new site. In this profile, the undamaged shells at the same depth are presumed to be approximately contemporaneous with the deposition of the clay and nodules, and thus to be indicative of the age of the deposit.

The multiple dates on shell and on hard and soft carbonate from the Tagra pit beside the White Nile (Table 1 and Figures 9b and 11) support the conclusion that hard sub-rounded carbonate nodules are often allochthonous and much older than the probable age of deposition of a sediment, whereas dates on soft carbonate are likely to be closer to the dates determined from mollusca. The date on the soft carbonate may reflect a minute component of ancient carbonate remobilised from the hard nodules.

The justification for accepting mollusca as the most reliable material available for dating is based on the lack of mineral alteration in the shells and on the modern dates (Table 1) obtained on several samples of shells from living mollusca from the White Nile and on one sample from the Gezira main canal fed by the Blue Nile. It is therefore improbable that mollusca incorporate ancient carbonate into their shells in this environment, despite the presence of some old carbonate in the fluvio-lacustrine sediments through which the rivers flow. We shall also see later in this chapter that dates on mollusca from central Sudan are congruent with palaeo-environmental evidence from elsewhere in Africa.

Concentrating now on the carbonate deposits along the Blue Nile, the ages range from Holocene to beyond the range of dating (Table 1) and we consider that both autochthonous and allochthonous carbonate are represented. Near Umm Dom Island the Blue Nile has intersected an extensive deposit of clay heavily impregnated with carbonate which bears no relationship to the deposits of the Blue Nile in the present or in the recent past (I-9289 and I-9290). This horizon (I-9289 and I-9290) of massive clayey calcrete extends well back from the western cliffed river edge and includes diffuse and nodular carbonate, the whole having the appearance of formation *in situ.* Even on the present cliff face which is subject to the addition of modern carbonate during Blue Nile floods, the carbonate is beyond the range of dating. We conclude that this carbonate-rich clay was laid down in Lake Pleistocene times or earlier and has been subject to little alteration since. The deposit has now been exhumed by the incision of the Blue Nile. The deposit is only a few metres below the level of the Gezira plain at its Khartoum end and it is not covered deeply by modern sediments. Not all carbonate-rich deposits presently forming low cliffs along the banks of the Blue Nile are as ancient as the Umm Dom deposits. Closer to Khartoum, dates of 11 560 ± 160 and 7 830 ± 120 BP (I-9287 and I-9288) were obtained from the cliff face exposed to the high levels of the river. In this case no subsurface sample was obtained back from the river bank as a check on deposition of modern carbonate by the river. We also suspect that deposits containing a mixture of modern and old carbon are being formed to-day on shoals of carbonate rock exposed in the bed of the Blue Nile at low water. On the western shore opposite Umm Dom Island these surfaces are encrusted with abundant algae which are capable of adding modern carbon and building up the deposit. They also, incidentally, shelter abundant colonies of the sponge

Stratospongilla bombayensis, which was first recorded from the Blue Nile at this site (Adamson & Williams 1973).

With respect to the validity of the volcanic ash dates previously discussed, it is important to note that the clay and ash from which the carbonate was obtained contained little or no obvious allochthonous carbonate. Further, the morphology of the dated carbonates was typical of autochthonous material. The abundant micro-fossils present in some of the ash deposits are also consistent with the waters carrying the ash being slow-moving, and in atmospheric equilibrium with respect to carbon dioxide. The carbonate used to date the ash certainly formed after the ash was deposited so that it should give a younger date than the ash deposit itself. On the other hand, contamination by inert carbon will increase the apparent age. Considering these opposite tendencies and the nature of the carbonate and the surrounding deposits, we consider that the ash was deposited towards the end of the Late Pleistocene and that the carbonate dates are close to the actual date.

The deposits along the Blue Nile also include rolled carbonate gravels, often cemented by later carbonates. These gravels may contain fossils and artefacts. The deposits recorded by Arkell in 1949 and by Berry et al. in 1962 (Whiteman 1971) at Singa and Abu Hugar are of this type. The Singa skull came from such a deposit. The radiocarbon date obtained by Whiteman (1971) of 17 300 ± 2 000 BP (I-712) on a tooth from a crocodile skull in the fluvially-bedded carbonate gravels at Abu Hugar was considered to be of doubtful value because of the probability of impregnation by young carbon. The site was close to low water level (as controlled by the Sennar Dam), more than 14 m below the top of the section, and exposed to annual wetting and drying by modern Blue Nile water. The age of this deposit may be far older than the probably unreliable radiocarbon date. The rolled carbonate gravels were probably derived from older carbonate-impregnated clays and silts as a result of fluvial reworking. Younger clays and silts overlie the carbonate gravels at Singa and Abu Hugar. These younger deposits contain varying amounts of carbonates, the erosion of which could contribute gravel to the bed load of the present Blue Nile. Were aggradation to take place in the present Blue Nile channel so that the present river-bed sediments became buried, the outcome would be a complex set of deposits in which the buried river gravels were a mixture of the basal gravels (reworked at least twice) with an input of much younger carbonate gravel from the overlying clays and silts. We conclude that nodules formed *in situ* throughout the various clays and silts become concentrated in coarse river-channel deposits when the clays and silts are eroded by lateral or vertical river action, and that any particular deposit of river-laid carbonate gravels in the Gezira may be a repository for carbonate of diverse ages.

Several important conclusions result from the study of carbonates:

1. Environmental conditions have been favourable for carbonate deposition in the Gezira sediments during much of the Late Pleistocene and probably earlier.

2. In some cases massive carbonate beds or calcretes have been formed suggesting the prolonged and high input of dissolved carbonate under condi-

tions suitable for its precipitation. Such conditions could include sluggish stream flow and wide dispersal of the water at the downslope end of a fan such as the Gezira.

3. Under less extreme conditions, enough carbonate has commonly been available for its deposition in most silts and clays, so that scattered *in situ* carbonate is usually present in most bodies of sediment.

4. Solid carbonate particles, usually in the gravel-size class, may be repeatedly reworked and they may become concentrated in massive fluvially-bedded deposits. Small allochthonous particles may also occur widely distributed in silts and clays such as in the Gezira clay soil. The age of such gravels could be variable from place to place.

5. Dissolved carbonate in Blue Nile water (at least in the main feeder canals coming from the Sennar Dam) and in White Nile water is in equilibrium with the present atmosphere as shown by our dates on modern mollusca. We assume that this was so in the past as well. Autochthonous carbonates precipitated directly from such sources are probably fairly suitable materials for dating, although contamination from some of the older carbonates is probably inevitable in most cases. Our results from the Tagra site give some basis for assessing this error (SUA-69B on soft autochthonous carbonate compared with the two shell dates SUA-68 and I-1485).

6. Deposits of ancient autochthonous carbonate are present very close to the surface of the Gezira alluvial fan at least at its Khartoum end. This evidence supports the conclusion reached when considering the volcanic ashes, that the surface of the Gezira fan is not of uniform age but is time-transgressive.

9.2.3 *Palaeochannels across the Gezira*

Palaeochannels on the surface of the Gezira plain were mapped in a preliminary fashion by Williams & Adamson (1973). They have also been briefly discussed in chapter 7. Taken in conjunction with the longitudinal section of the Gezira visible in the Blue Nile kerrib, they are a major source of evidence about the later Pleistocene and Holocene history of the Blue Nile and its fan. A revised map showing in more detail the location of palaeochannels of the Blue Nile, Rahad and Dinder Rivers is shown in Figure 3. The palaeochannels, like the modern channels, were confined between the basement rocks of the Manaqil Ridge on the west and outcrops such as Jebel Fau on the east. Between J. Moya and J. Fau, near-surface rocks such as the bar at Sennar have probably exerted a limit on incision or lateral wandering by the rivers crossing this region. Upstream of Sennar and downstream of Abu Hugar are well preserved major palaeochannels on the western side of the Blue Nile. Palaeochannels also occur between the Blue Nile, Dinder, and Rahad Rivers in this region but they have not been well mapped. From about 30 km north of Sennar to Wad Medani there are a series of high-level meander cut-offs or oxbows on both sides of the Blue Nile, now abandoned as a result of incision by the Blue Nile. Major palaeochannels cross the Gezira in a generally north-westerly direction from north of Sennar towards the White Nile. Al-

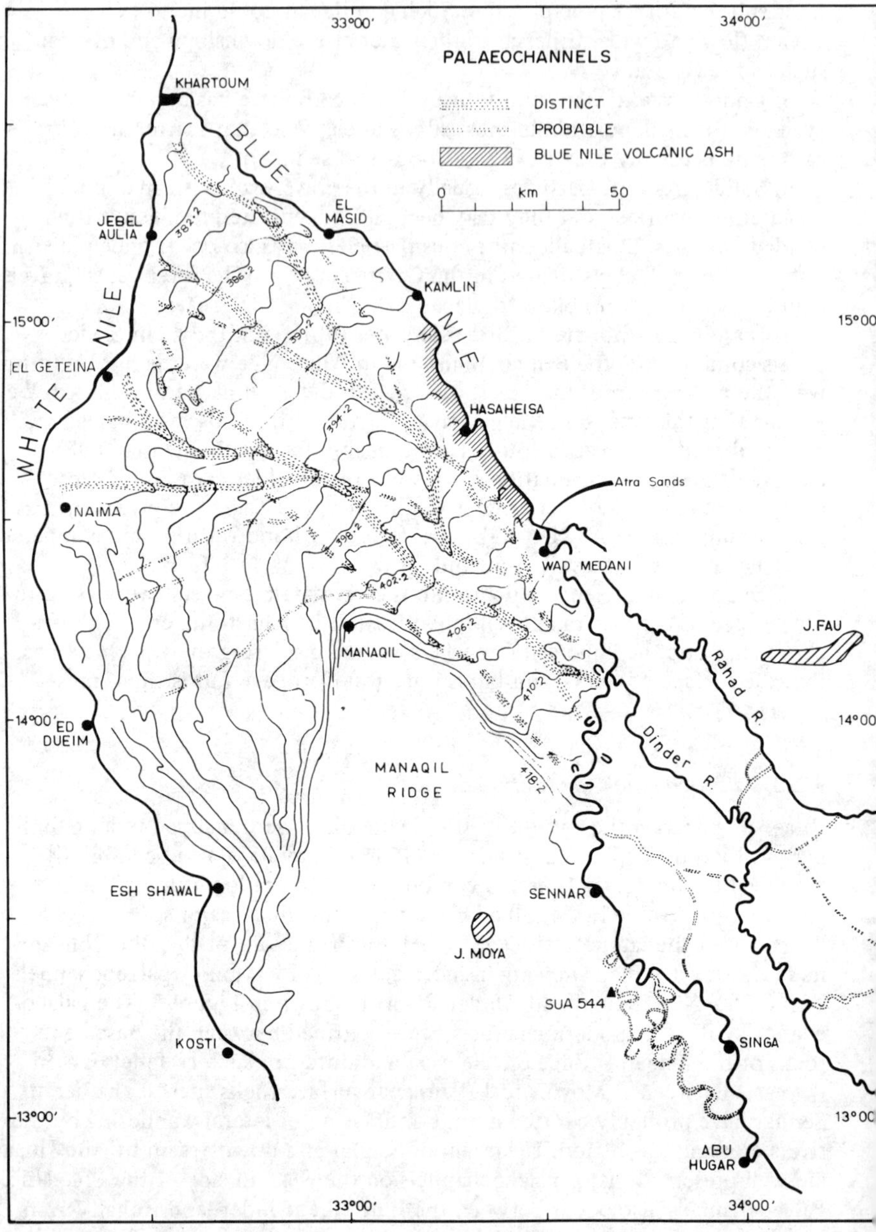

Figure 9.3. Distribution of palaeochannels on the present surface of the Gezira plain. Palaeochannels are also shown west of the Blue Nile between Sennar and Abu Hugar. Between the Blue Nile and the Rahad River palaeochannels have not been mapped accurately. High-level oxbows on both sides of the Blue Nile are shown between Sennar and

though not marked as such on Figure 3, the vicinity of the present Blue Nile was probably also a major palaeochannel.

A reconstruction of the late Pleistocene-early Holocene fluvial network is shown in Figure 4 and compared with the present river pattern. Not all the palaeochannels shown as connected in Figure 4 necessarily functioned simultaneously or were of equal importance. However if such a drainage network was not deeply incised, water would probably flow along most of the channels in the flood season. With high river discharge, water would also spill across the intervening flat country by overbank flooding, and leave extensive pools and swamps in the less active channels well beyond the end of the flood season or even throughout the year. With lower discharge, seasonal flooding of the channels would have been followed by annual drying out of most of the channels. During periods of change in discharge or load, instability in the network would occur leading to the creation of many rapidly changing distributaries or the opposite trend towards the consolidation of flow into one or a few major channels.

If the radiocarbon dates associated with the volcanic ash channels are accepted, then around 25 000 BP there was probably a network of narrow relatively straight interconnected channels distributing the flood peak from the main channels widely across the fan surface, extensive overbank flooding depositing the clays which flank the minor channels, and, overall, a relatively unincised drainage network. At the lower or northern end of the fan this well-distributed flow would spill into the White Nile trough and evaporate on the interfluves.

On the Gezira there is a dearth of dates for the period c. 12 000 to c. 20 000 BP. The interval c. 12 500 to c. 17 000 BP is one of generally low lake levels and of relative dryness across a wide belt of Africa including the catchment areas of the Nile basin (Chapter 2 and Figure 21). There can be little doubt that the catchment of the Blue Nile and its tributaries supplied substantially less water at this time than it has done during most if not all of the Holocene period. If lakes were generally low or dry during this interval in Arabia, the Ethiopian rift, Kenya and Chad, then it is likely that the discharge of the Blue Nile and its tributaries was on average also much lower during this interval than either before or since. The Blue Nile was also probably even more seasonal than it is today. The Gash and Baraka Rivers today may be minor examples of the type of fluvial regime of the Blue Nile at least during the most arid periods. The peak flow was still sufficient to carry a seasonal flood across the Gezira and down the main Nile. In the dry season there may have been very little permanent water on the Gezira and few suitable habitats for mollusca whether they be permanent water, amphibious or

---

Wad Medani. The generalised distribution of volcanic ash deposits exposed in the eroded land (kerrib) along the western side of the Blue Nile is indicated. The Atra sand pits, and the location of the dated Nile oyster (SUA-544:11 330 ± 150 BP) from the bank of a palaeochannel between Sennar and Singa, are also shown.

terrestrial species. Further evidence about the fluvial and environmental conditions on the Gezira during this time is urgently needed.

Dates between 11 000 and 12 000 BP have been obtained on shells from three sites near the Blue Nile (SUA-544, 604 and 214). These three dates, and two from near Esh Shawal near the White Nile (SUA-75 and I-1486) are informative. They correspond in time to a rise in lake levels that occurred across Africa, and to a period in which discharge by the Blue Nile must have increased. Dramatic related events include the overflow of Lake Victoria to feed the White Nile (Kendall 1969), and a change in behaviour of the main Nile in Egypt, including a brief period of exceptionally high flooding c.

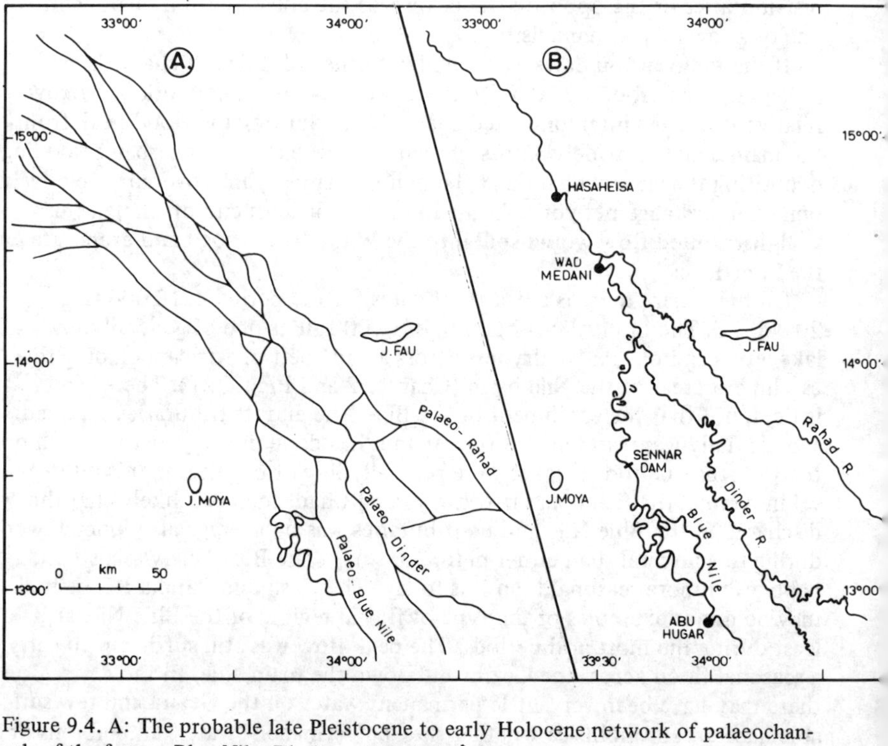

Figure 9.4. A: The probable late Pleistocene to early Holocene network of palaeochannels of the former Blue Nile, Dinder and Rahad drainage system. The palaeochannels are shown in their presumed correct positions but no attempt has been made to show their degree of meandering, except in the case of the meandering channels between Sennar and Abu Hugar where meander scars have been very well preserved. On the southern part of the main Gezira plain, channel form is usually obscured beneath superficial clays. It is not implied that all the palaeochannels were functional simultaneously or were of equal importance.

B: Present drainage pattern of the Blue Nile, Dinder and Rahad Rivers.

11 500 to 12 000 years ago (Butzer & Hansen 1968:115-116, 328; Butzer 1980). This period also seems to have been a time of dramatic change on a global scale, involving rapid melting of polar ice and a consequent influx of fresh water to the world's oceans, and probably rapid temperature changes (Berger et al. 1977).

The date of 11 330 ± 150 BP (SUA-544) on Nile oyster comes from the bank of a major palaeochannel of the Blue Nile south of Sennar (Figure 3 and Table 1). This channel forms an extension of the Khor Umm Ish complex of palaeochannels. The site was beside the main Sennar-Singa road on the northern bank of the palaeochannel about 50 cm below the present surface. Shell, bone and burnt clay were exposed in scattered occurrences where the road had made a very shallow cutting into the dusty carbonate-rich sediments of the bank. The palaeochannel is a broad, very shallow, elongated depression meandering across the dark grey-brown clay plain. The channel banks, in contrast to the plain, are pale brown in colour due to powdery carbonate. The channel was undoubtedly deeper and occupied by permanent water when the oysters were collected and carried up onto the channel bank. The channel probably remained active long after 11 000 years ago but became an unsuitable environment for the oysters as a result of rapid aggradation and shallowing of the bed by silt. Prior to 11 000 to 12 000 years ago, flow in the channel may have been too seasonal to support the oysters. Ultimately, in the Holocene, the whole Khor Umm Ish system became defunct, choked with silt and abandoned in favour of the present channel of the Blue Nile. Downcutting by the present Blue Nile, although limited by the rock bar at Sennar, has also contributed to the drying up of the palaeochannels. The Khor Umm Ish system had a fairly long period of fluvial activity in the late Pleistocene and Holocene to judge from its multiple channels and its extensive series of meander scars.

On the Gezira fan itself, north of Sennar, high-level abandoned meanders flank both sides of the Blue Nile (Figure 3). These oxbows represent relics of a meandering and relatively unincised Blue Nile drainage network which was probably functional at the same time as the Khor Umm Ish system. Several oxbows are shown in more detail in Figure 5, and details of the elevation and relief of three oxbows, at Hag Abdulla, opposite the confluence with the Dinder River, and at Barakat, are shown in Table 2. Their elevation is above flood level of the present river. The beds of these meanders are now covered with silt and they are now certainly much shallower than formerly. Augering into the Barakat palaeochannel revealed fine fluvial sands with a typical Blue Nile suite of minerals beneath 3.3 m of clay, silt, and clayey sand. As with the Khor Umm Ish channel, these meanders became lined with finer sediment as they were gradually abandoned.

Palaeochannels are also revealed by shallow closed depressions on the Gezira plain (Figure 5 and Table 2). These depressions form part of the drainage network which crosses the Gezira and which is associated with sandy bed-load deposits further down the fan (Figures 3 and 4 and Figure 3 in chapter 7). After the abandonment of the channels by the Blue Nile, these depressions and channels would have been important locally as seasonal swamps

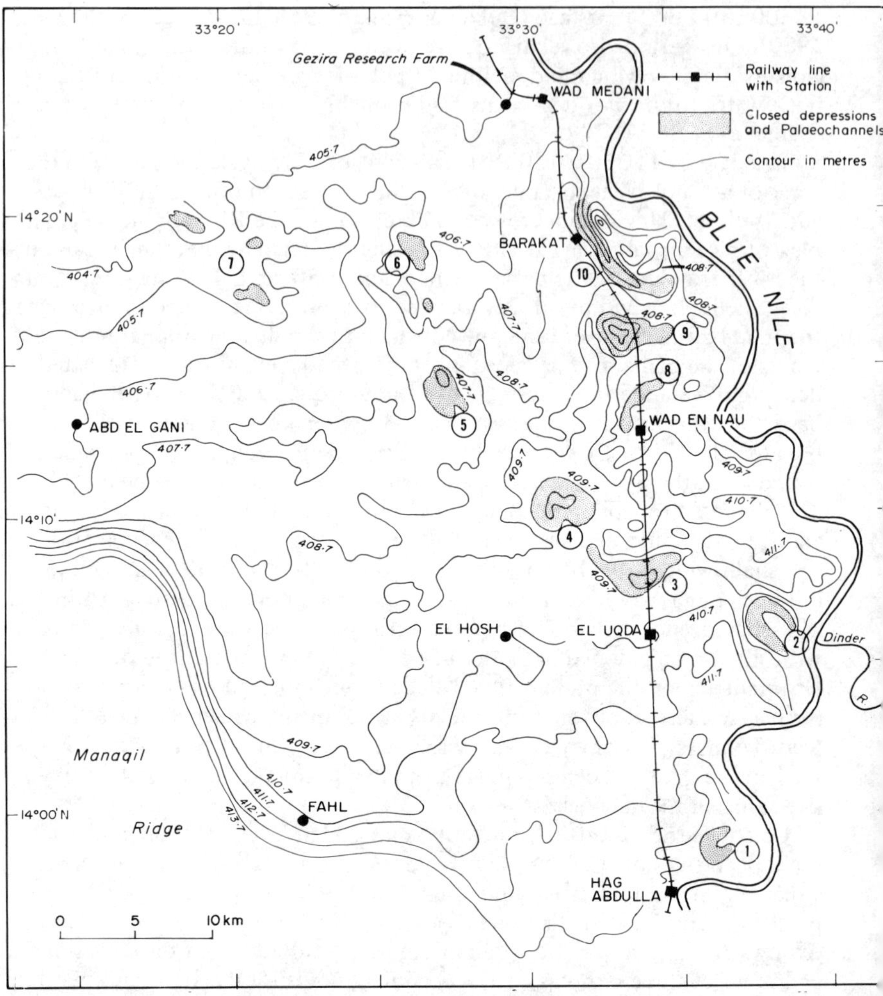

Figure 9.5. Closed depressions and high-level oxbows on the southern Gezira plain. This map shows, in more detail than Figure 9.3, the low-relief valleys on the clay plain. The numbering of the depressions and oxbows matches that in Table 9.2.

and ponds. They would have supplied water on the Gezira surface well away from the Blue and White Nile for several months during and after the summer rains. Today some of the closed depressions contain water in a good wet year and these depressions have undoubtedly been important for game and for hunters until modern times.

The bed load sands and gravels of a vigorous palaeochannel are exposed

Table 9.2. Depressions and oxbows associated with the Blue Nile and Gezira Plain south of Wad Medani, central Sudan.

Location	Elevation of floor of depression to nearer 0.5 m (m)*	Height of floor of depression above flood level of Blue Nile (m)**	Height of Gezira Plain above floor of depression (m)
1. Hag Abdulla 13°59′N 33°36′E	411.7	6.8	2
2. Dinder-Blue Nile Confluence 14°06′N 33°38′E	403.7	1.0	8
3. El Uqda kilo 205 on railway 14°08′N 33°34′E	408.7	6.0	2
4. 14°10′N 33°31′E	408.7	–	1.5
5. 14°14′30″N 33°27′30″E	406.7	–	1.5
6. 14°19′N 33°26′E	405.2	–	1.5
7. 14°18′N 33°20′E	404-405	–	0.5
8. Wad En Nau 14°14′N 33°34′E	407.7	7.2	2
9. kilo 194 on railway 14°16′N 33°33′30″E	406.2	5.9	3-3.5
10. Barakat 14°18′N 33°33′E	405.7	7.0	2.5-3

* metres above mean sea level at Alexandria, assuming Khartoum gauge is 360.00 m (actual level is 362.26 m)

** estimated by assuming a uniform slope on the river water level between Hag Abdulla and Wad Medani gauges and by using the maximum 10-day mean river heights recorded at Hag Abdulla and Wad Medani gauges up to 1927 (Hurst & Phillips 3:19).

in the extensive Atra sand quarry, in the bend of the Blue Nile opposite its junction with the Rahad River (Figure 3). The quarry is about 8 km north of Wad Medani and 1 to 2 km from the southern end of the series of volcanic ash deposits beginning at Wad el Magdoub. The relation between the sand pits, the Gezira plain and the Blue Nile, is shown in Figure 6a, the deposit occurring in the kerrib about half way between the elevation of the plain and the present maximum flood height of the Blue Nile. Figure 6b shows current bedded sands and gravels typical of an active river-bed overlain by horizontally bedded sands and gravels. Carbonate nodules are common in the gravel fraction. In other sections, finely-laminated silts occur near the

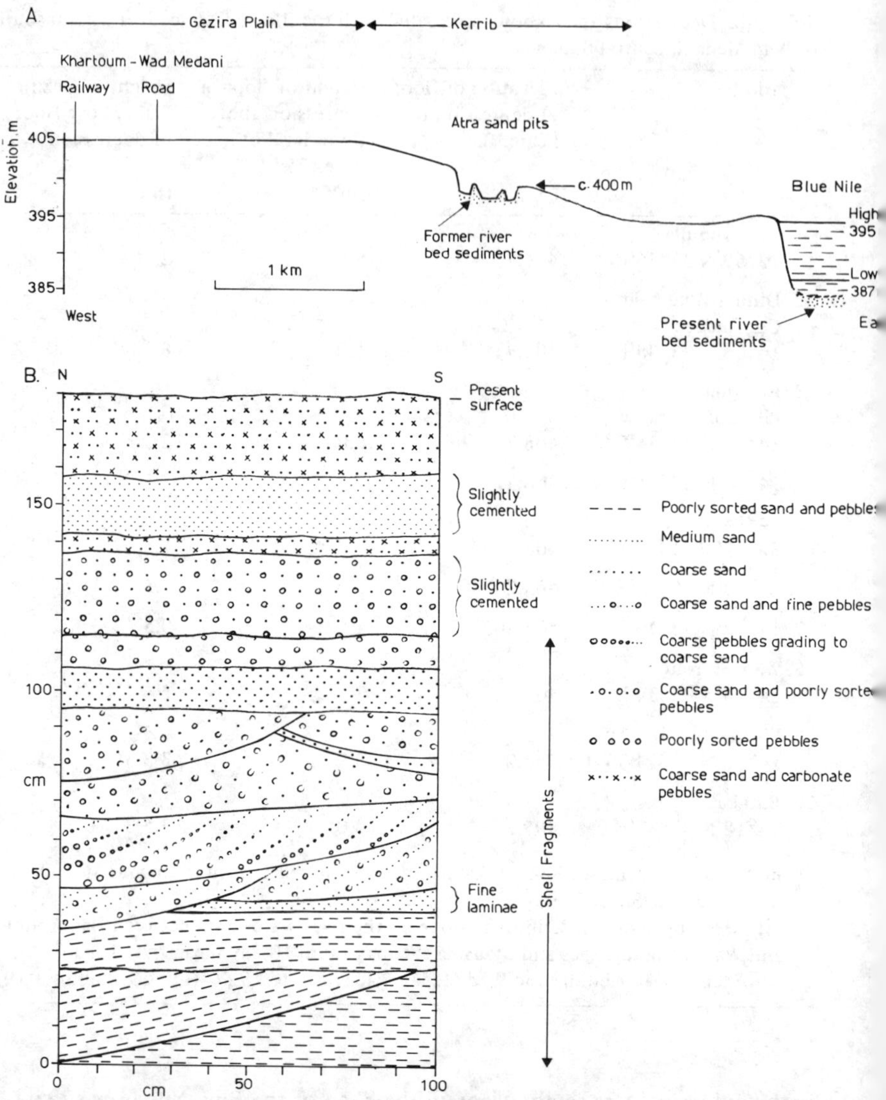

Figure 9.6. A: Section from the Gezira plain to the Blue Nile through the sand pits at Atra (see Figure 9.3 for location).

B: Section of fluviatile sands and gravels exposed in the sand pits at Atra. The shell sample dated 11 800 ± 195 BP (SUA-604) came from current-bedded sands and fine gravels similar to those shown in this section.

top of the sequence. Whole shells and shell fragments are common in the lower sands and gravels, with dense accumulations of intact shells of species such as *Cleopatra* occasionally occurring in the current bedded sands. Such a deposit gave an age of 11 800 ± 195 BP (SUA-604), undoubtedly indicative of the time of deposition. This bed load deposit occurs at about the same level as the volcanic ash deposits but it is probably considerably younger. If the average slope of the Gezira is taken into account, it also occurs at about the same relative elevation as the sands which occur beneath the superficial clay and silt in the Barakat palaeochannel. We consider that the Atra river channel deposits are part of the palaeochannel network that was active at the close of the Pleistocene and that they are related to the abandoned oxbows and channels already described.

The Atra channel is probably a relic of a prominent palaeochannel which followed closely the present course of the Blue Nile. Further evidence of major former channels paralleling the present course of the Blue Nile occurs at Hasaheisa (Chapter 7, Figure 8) where the scar of an abandoned high-level river channel is present. The sands now being blown out of the Blue Nile to form the active high dunes at Hasaheisa may come from palaeochannel deposits exhumed by the present incised river as well as from the present bed load carried from upstream.

The third important date in this region comes from the pit dug on the Gezira Research Station at Wad Medani (Figure 8b). The date on mollusca, at a depth of 160-170 cm, of 11 975 ± 260 BP (SUA-214) indicates that 1.6 m of clay has been deposited at this site since that time. The shells occurred in the grey clay which commonly underlies the dark brown surface clay of the typical profile of the Gezira clay soil. We conclude that the grey clay was being deposited around 11 000 to 12 000 years ago and that the overlying brown clay is early Holocene or younger.

The river channel carrying the Atra sands 11 000 to 12 000 years ago lay close to the surface of the Gezira plain, particularly if the upper 1 to 4 m of clay on the plain was not present at that time. Figure 6 shows that the river bed sediments at Atra were not more than 4 m below the contemporary surface of the surrounding plain. A similar situation occurred with the Barakat and Hag Abdulla oxbows and probably also with the Khor Umm Ish palaeochannels. Such shallow unincised channels must have overflowed and inundated the countryside when flood discharges were high. The palaeochannels crossing the Gezira (Figures 3 and 4) would have behaved similarly. We have already concluded that the flood discharge of the Blue Nile increased greatly at about this period, making inevitable major seasonal inundation of large areas of the Gezira from the network of shallow river channels inherited from the previous more arid interval.

The distribution of mollusca in the upper 2 m of Gezira clay is consistent with this interpretation. Some of Tothill's classic data (Tothill 1946) are presented in generalised form in Figures 7 and 8a. Figure 7 shows the distribution of sub-fossil shells of three species of mollusca across the Gezira, and Figure 8 shows how the distribution of shells varies down the soil profile. In both cases, permanent-water, amphibious, and terrestrial species are shown. Figure

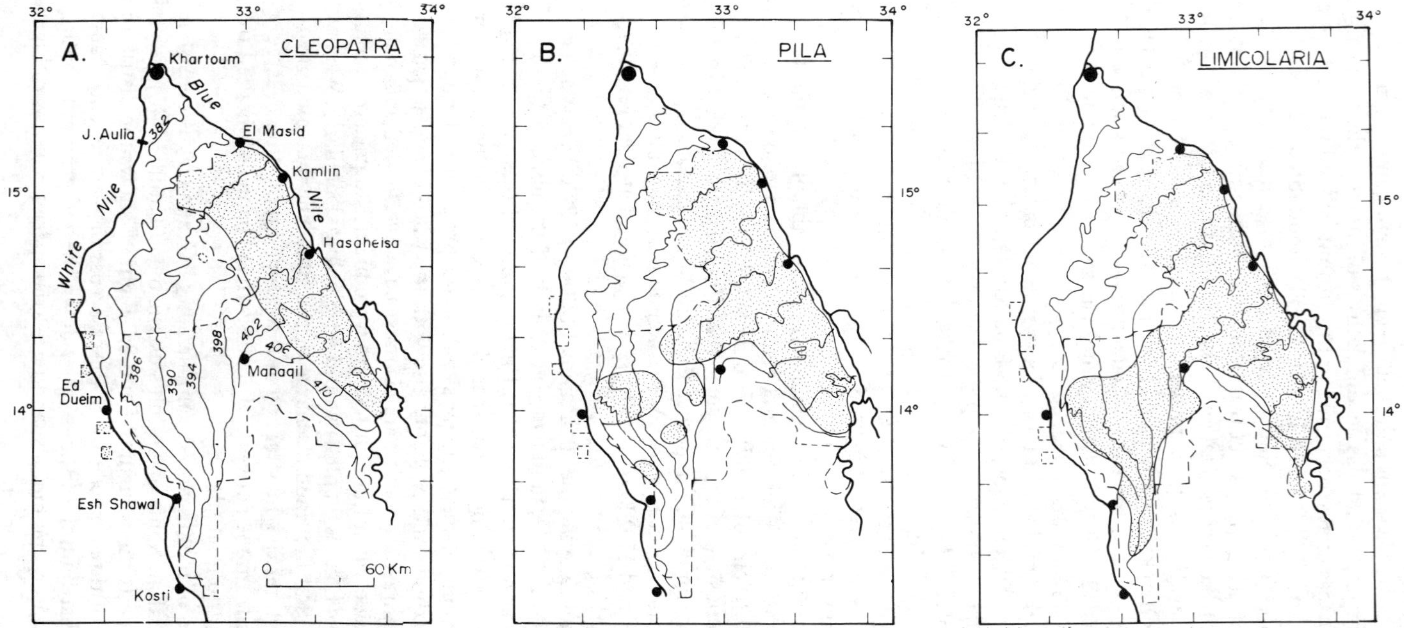

Figure 9.7. Distribution of sub-fossil mollusca in the upper two metres of the soils of the Gezira. The maps have been redrawn and generalised from the original data presented by Tothill (1946). The dashed lines represent the boundary of the area surveyed, not the limits of distribution of the mollusca.

A: *Cleopatra bulimoides* (Oliv.), a species dependent on permanent water.

B: *Pila wernei* (Phil.) *(Ampullaria)*, an amphibious species that can aestivate in moist soil during a dry phase between seasonal flooding.

C: *Limicolaria flammata* (Caill.), a land snail that is intolerant of flooding and now lives in areas with a minimum rainfall of c. 350 to 400 mm (compare with the present rainfall on the Gezira shown in Figure 7.15).

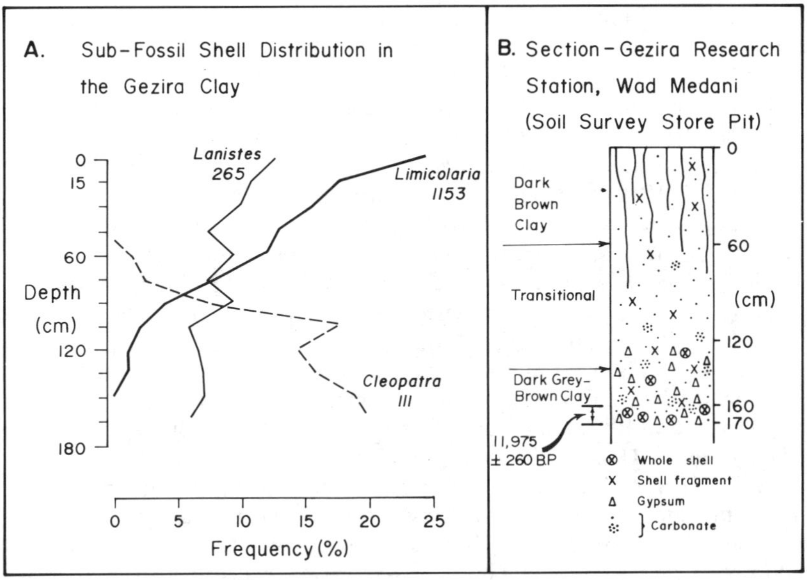

Figure 9.8. A: The vertical distribution of mollusca in the Gezira clay soils. Data of Tothill (1946) collated and plotted in graphical form. The numbers represent the total number of individual shells of each species. *Lanistes carinatus* (Oliv.) is an amphibious species; for other species, refer to legend of Figure 9.7.

B: Soil section, Gezira clay soil, foundation pit for the Soil Survey Store building, Gezira Agricultural Research Station, Wad Medani. Radiocarbon dates on shell (*Cleopatra* and *Lanistes* SUA-214; 11 975 ± 260 BP) and on small, smooth, grey-black, hard carbonate concretions (SUA-386, 27 370 ± 800 BP) came from a depth of 160 to 170 cm in this profile.

8b illustrates a typical Gezira soil profile in the pit which supplied the hard carbonate nodules and shells for the SUA-386 and SUA-214 radiocarbon samples.

Cleopatra is a mollusc requiring permanent water. Its distribution across the Gezira plain (Figure 7a) and its vertical distribution in the Gezira soil (Figure 8a), are both consistent with extensive areas of permanent water and seasonal flooding along the eastern side of the Gezira. Tothill (1946) pointed out that all ages of this mollusc were represented in the soil samples, suggesting temporary flooding each year followed by drying out of the water and mass death of all age groups. More restricted areas of permanent water would have remained after the passage of the flood. Amphibious mollusca (Figures 7 and 8) such as *Pila* and *Lanistes* would survive in the mud which would remain wet beneath a dry surface crust until the next flood. Being amphibious, they would also have a wider distribution into less wet areas further down the Gezira fan as shown by Figure 7. *Limicolaria,* being terrestrial, would find conditions too wet at this time. The wide distribution of *Limicolaria* shown in Figure 7 extends beyond the limits of Blue Nile flooding up onto the Manaqil Ridge. This, and its abundance towards the top of the profile shows that *Limicolaria* occurred in this area somewhat later, towards the end of clay deposition on the Gezira during early to middle Holocene times, when the palaeochannels were becoming defunct but when the local rainfall was substantially higher than at present. We date this interval at between 5 000 and 8 500 BP with a possibly drier phase between about 6 000 and 6 500 BP (Figure 21). The very wet conditions on the Gezira which were suitable for *Cleopatra* probably covered a strip at least 180 km in length by almost 50 km in width on the eastern side of the Gezira during the period of about 11 000 to 12 000 BP.

The previous discussion implies that the volume of water that could be accommodated within the various palaeochannels at bankfull flow at about 11 000 to 12 000 BP was substantially less than the total volume of water flowing onto the Gezira. The excess water covered large areas of the plain, brought in a suspended load of clay particles, and allowed the growth of permanent-water mollusca. In this connection it is worth noting that even if the whole of the three months' flood flow of the present Blue Nile was simultaneously spread as a standing body of water over this part of the Gezira, it would cover it to a depth of the order of only 1 m. The suggestion that a substantial proportion of the flow of the late Pleistocene peak floods was spread out across the Gezira is therefore plausible. The body of water was of course not a standing one, and channels also carried a substantial through flow, so that there is no particular difficulty in getting a sufficient combined discharge from the Blue and White Nile into the main Nile to account for the unusually high flood levels in the Egyptian Nile previously mentioned. We shall see later that events in the White Nile could have made an important contribution to short intervals of dramatically high flood levels in the main Nile.

9.2.4 *Beheading of the palaeochannels*

Flow of huge quantities of water across the Gezira eventually declined during the Holocene until all the channels became defunct and all the Blue Nile water became channelled down its present trough. The distribution of mollusca in the soils of the Gezira reflect this trend to the present-day drainage pattern. The mollusca also attest to the changes in the local climate which have contributed to the drying out of the Gezira. Figure 8a shows that the land snail *Limicolaria* becomes common towards the top of the Gezira soil as the permanent water *Cleopatra* disappears. Conditions remained favourable for the amphibious *Lanistes* throughout deposition of the Gezira clay. This vertical succession is also a time sequence indicating a progressive reduction in the volume and the duration of flooding of the Gezira from the Blue Nile.

There is no evidence to suggest that low Blue Nile discharges occurred during the early to middle Holocene which could have accounted for the drying out of the Gezira palaeochannels, particularly as these channels were previously at least seasonally active during the long period of low discharge in the arid terminal Pleistocene interval before 12 000 BP. The explanation lies in a switch of fluvial regime from one characterised by multiple interconnected aggrading channels to one with a single major incising channel. The Gezira palaeochannels became progressively beheaded between Sennar and Wad Medani. During the transition from aggradation to incision the rate of deposition of the final metre or so of silt and clay on the Gezira probably declined exponentially. At the time when *Cleopatra* thrived and the dark grey clay was accumulating, massive water inflows were probably associated with massive silt deposition. The mean annual rate of deposition would have been high, probably an order of magnitude higher than the closing rate.

The cause of the change in fluvial regime remains obscure because it could be caused by changes in both discharge and load, or by either alone, changes which can only be analysed roughly on the basis of our present knowledge of both fluvial dynamics and events in the headwaters of the Blue Nile. It is likely, however, that the greatly increased flood discharge, which we consider to have occurred since about 12 000 BP, was one major cause of the change. Working out the causes and detailed sequence of the change in fluvial behaviour remains one of the major problems connected with the late Quaternary evolution of the Blue Nile.

In general terms, the important change was from aggradation to incision. This resulted in the gradual dominance of one or a few channels which incised faster than the others so that flow was progressively concentrated and incision accelerated. In the early stages of this process all the original channels would continue to receive water carrying suspended load but they would be deprived of bed load. Progressively less and less of the flood flow would be diverted into them as the main channel incised further and their own channels became lined with silt. Ultimately the minor distributaries would cease overbank flooding so that deposition on the surrounding plain ceased. They still carried water from the peak of the flood in the main channel, but this only contributed silts and clays for their further aggradation. Eventually the subsidiary channels were deprived of all water except run-off from local rain.

The palaeochannel crossing the Gezira plain in a north-easterly direction from the confluence of the Blue Nile and the Dinder River (Figures 3 and 5) is one of the most prominent, and it may have been one of the last to become abandoned. This palaeochannel starts close to a particularly deep ox-bow on the western side of the Blue Nile (Table 2), and follows a continuous belt of low country including a succession of closed depressions and of shallow, branching valleys to the White Nile in the vicinity of El Geteina, Shabona and Naima. The alignment of this channel at its southern end looks like a continuation of the Dinder River. Further north on the Gezira, the alignment of another major palaeochannel running north-east through the qoz deposits of Wad Es Said and Qoz Kabaro (Figure 3) matches that of the Rahad River. It is possible that, during the fluvial readjustment at the end of the Pleistocene and in the early Holocene, the dominant channel shifted laterally in a stepwise fashion until it became firmly established near the eastern margin of the fan in the present channel of the Blue Nile. Alternatively, the present position of

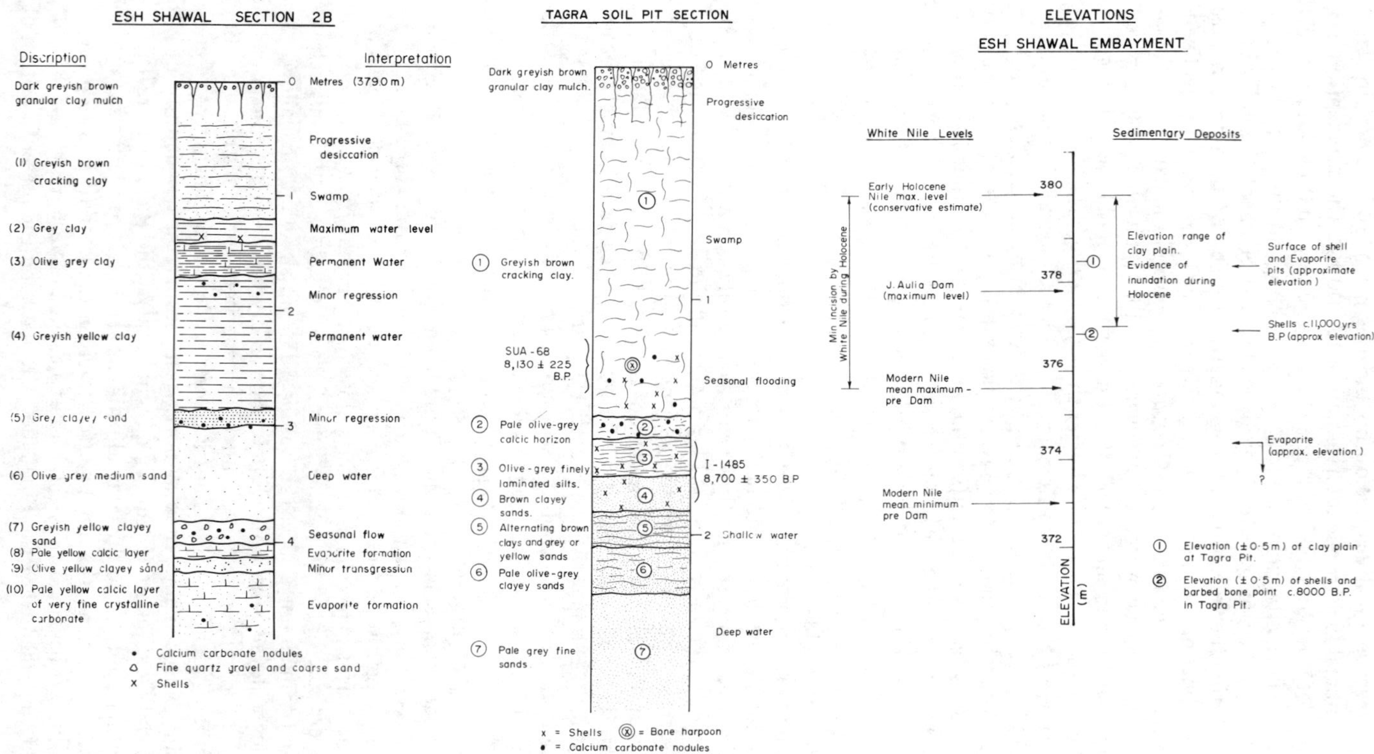

Figure 9.9. A: Shawal Profile.
B: Tagra Profile.
C: Elevations, Esh Shawal Embayment.

the Blue Nile downstream of the Rahad River has long been occupied by a major channel carrying water from the upper Blue Nile and Dinder as well as from the Rahad. Deepening of this major channel could then have beheaded the other channels running across the Gezira and deprived them of Blue Nile and Dinder water. In either case, because the volume of the Blue Nile flood would always have been far in excess of that of the Rahad and Dinder floods, even in periods of low discharge, it is probable in the case of an unincised drainage network that Blue Nile water forced its way into the palaeo-Rahad and palaeo-Dinder distributaries through the lateral interconnecting channels shown in Figure 4.

The dates clustering between 11 000 and 12 000 BP show that the river channels were not deeply incised into the Gezira at that time and that over-bank flooding occurred extensively across the Gezira. Archaeological evidence from the site of the Early Khartoum Culture (Arkell 1949) shows that the levels attained by Nile floods at that time in Khartoum were substantially higher than they are now. A flood level of about 10 m higher than present was indicated by Arkell. We now know that the age of the Early Khartoum Culture was c. 7 000 to 9 000 BP (SUA-68, I-1485, SUA-298, and Adamson et al. 1974), so that incision at Khartoum had not progressed far by that time. However, Blue Nile discharges may then have been higher than at present resulting in higher flood rises. This could have counteracted the effect of some incision. This period was locally wetter than at present (Figure 21, and Williams et al. 1974) and the moister climate probably influenced the Blue Nile catchment as well.

By 6 890 ± 115 BP (SUA-215A) Blue Nile water was still flowing in the palaeochannel passing the present villages of Wad Es Said and Qoz Kabaro (Chapter 7, Figure 3). This date was obtained on shell of the amphibious mollusc *Pila* located *in situ* near the surface of fluvially-bedded sands near Wad Es Said. Bone and artefacts also occurred nearby. The deposit of shell was undisturbed by later fluvial activity nor was it buried under later deposits. This date may indicate the waning of activity in this major palaeochannel. The valley would however continue to collect local run-off and could have supported swamps in favoured localities. Wind-blown red sands forming source-bordering dunes also occur in the vicinity, particularly near Qoz Kabaro, but they are quite distinct from the fluvial sands.

By 6 000 to 7 000 years ago the Blue Nile was probably firmly established in its present location, sufficiently incised to have deprived some Gezira palaeochannels of their water, but still supplying water to the deepest ones at the peak of its flood. We cannot date accurately the last flow of Blue Nile water in the last active Gezira channel, but it probably occurred within the period 5 000 to 3 000 BP. Major flooding of the Gezira ceased much earlier.

9.3 THE WHITE NILE IN THE LATE PLEISTOCENE AND HOLOCENE

The White Nile is a straight, shallow and remarkably stable river which flows at the very low flood gradient of 1 cm per km in an almost horizontal trough between Kosti and Khartoum. In its lower course it is confined between bedrock outcrops and two enormous low-angle alluvial fans, the fan of the Khor Abu Habl, and the Gezira fan of the Blue Nile (Chapter 7, Figure 1). The White Nile is also constrained by the Blue Nile directly, in that before construction of the Jebel Aulia Dam the flood level of the Blue Nile controlled the flood level of the White Nile for several hundred kilometres upstream of their confluence. This important control on the level of the White Nile has undoubtedly operated in the past. It is for this reason that we dealt with the history of the Blue Nile before turning to that of the White Nile.

9.3.1 *The seasonal White Nile in the Late Pleistocene*

Three independent lines of evidence show that the White Nile was a highly seasonal river in the past, in remarkable contrast to its present steady regime described in chapter 7.

Firstly, cores from the bed of Lake Victoria show that this was a closed lake before about 12 000 years ago for an unknown length of time (Kendall 1969). The earlier history of overflow from Lake Victoria into the White Nile is unknown, but the presence of evaporites in the White Nile trough (Williams & Adamson 1974) shows that overflow was at best episodic. When the White Nile was deprived of the large outflow and stabilising effect of Lake Victoria, it would have been highly seasonal with a negligible low season flow. This effect would have been accentuated because the extensive Sudd swamps which also smooth out seasonal fluctuations in flow would have shrunk or, more probably, disappeared. This time interval was one of aridity not only in the Lake Victoria basin, but also across intertropical Africa (chapter 9) so that input of water to the White Nile from all its sources would be reduced well below present levels to make them more seasonal than at present. We consider that low season flow in the lower White Nile may have ceased altogether during the most arid intervals in the few thousand years before 12 000 BP (Williams & Adamson 1976).

Extreme seasonality and intense evaporation are also indicated by the occurrence near Esh Shawal of deposits of very pure microcrystalline calcite and dolomite (Figures 9 and 10) well below permanent water mollusca dated at 11 250 ± 220 BP (SUA-75) and 11 300 ± 400 BP (I-1486). These beds of evaporites were interpreted by Williams & Adamson (1974) as dating from the arid period before about 12 000 BP. This estimate of the age of the evaporites was modified (Williams & Adamson 1976) in the light of a series of radiocarbon dates on the microcrystalline carbonates, the revised ages of which range from 25 600 to in excess of 40 000 BP (SUA-837, 381, 382). The evaporites predate the terminal Pleistocene arid phase and they may reflect extreme desiccation associated with one of the earlier late Pleistocene arid phases. Their antiquity is also indicated by the occurrence of several intervals of fluviatile sedimentation between evaporite deposition at a depth of about 4 m, and deposition of the shell-bearing green clays which date from a little over 11 000 BP. Figure 9 indicates a complex sequence of deposits and events. From the base upwards section 2B comprises over 1 m

Figure 9.10. Scanning electron microscope photographs of evaporites from the Esh Shawal Embayment, White Nile. Locations and sections are shown in Figures 9.15 and 9.16. Scale bars equal 10 micrometres.

A: Microcrystalline high magnesian calcite from Trench ES1 at a depth of 380 cm below the present ground surface. The deposit was mainly composed of individual spindle-shaped calcite crystals. The purity of the evaporite deposit is shown by the photo. Material as sampled was metal coated and photographed without any prior treatment.

B: Detail of a calcite crystal from Trench ES1.

Figure 9.10 (continued). Scanning electron microscope photographs of evaporites from the Esh Shawal Embayment, White Nile. Locations and sections are shown in Figures 9.15 and 9.16. Scale bars equal 10 micrometres.

C: Microcrystalline high magnesian calcite and dolomite from Trench ES3 at a depth of 410 cm below present ground surface. The deposit was mainly composed of individual calcite and dolomite crystals. This sample, as with A, was not cleaned or purified in any way before examination.

D: Detail of a dolomite crystal from Trench ES3. Electron microscopy by Dr C.Nockolds, Electron Microscope Unit, Sydney University.

of evaporite beds (full depth not penetrated); an upward-fining alluvial sequence of gravelly to clayey sand; a metre of clay whose surface was subject to soil forming processes; lacustrine olive-grey clays containing the dated permanent water mollusca; and grey-brown cracking clays forming the present surface which is now exposed to pedogenesis. Two holes which penetrated below 3.5 m showed the occurrence of evaporites, one containing only microcrystalline high-magnesian calcite, and the second a mixture of high-magnesian calcite and dolomite. Most of our holes and pits in this region have only penetrated to about 2 m so that the lateral extent of the evaporites is unknown. A survey of their extent and thickness is needed because of their implications for water and soil salinity (chapter 7) and because of their palaeogeographic importance.

The third line of evidence for highly seasonal flow of the river in the past is the occurrence of extensive sand dunes in the floor of the White Nile trough, particularly between Ed Dueim and El Geteina. These have been described in chapter 7 where they are illustrated by Figures 7, 8 and 9. The slight activity of some of them today is due to recent overgrazing. They are presently flanked and partly buried by alluvial clays and they show no sign of being contemporary structures. For example they are not analogous to the source bordering dunes now being formed beside the Blue Nile at Hasaheisa. They have been in existence, and their surface has been relatively stable, for at least 7 000 years (SUA-298). They are flanked by heavy clays containing mollusca at a depth of 1.5 to 2 m (Figure 7a). The same species of permanent water mollusca, at the same depth in similar types of alluvial clays have been dated at Tagra and at Esh Shawal as about 8 000 and 11 000 years old respectively (SUA-68, I-1485, SUA-75, I-1486). The sand dunes predate 11 000 BP. They have also been sculptured by wave and river action (chapter 7), which was probably at its peak soon after Lake Victoria began supplying a greatly increased and steady discharge into the White Nile after about 12 000 BP. On the other hand, they do not have the appearance of ancient features and the obvious influences that have impinged upon them date from the Holocene. We conclude that the sand dunes within the White Nile trough originated as source bordering dunes during the Lake Pleistocene arid phase preceding 12 000 BP. The sand source was the bed load of the seasonally fluctuating river and ancient sand deposits exposed to wind action at low river level. Such aeolian activity is difficult to imagine if the White Nile had a steady flow, as at present.

During the terminal Pleistocene arid period the palaeochannels of the Blue Nile flowing across the Gezira carried sand which was dumped into the White Nile trough, particularly between Jebel Aulia and Naima. This very large volume of sand added to that from the White Nile. As Williams (1966) pointed out, there is no need to invoke aeolian transport of sand from Kordofan. The more recent evidence confirms the fluvial origin of the White Nile and Gezira sands. Similarly, the sand which impinges on the White Nile from the west (chapter 7, Figures 7 and 9) was deposited by the distributaries of the Khor Abu Habl on its fan and then locally reworked by wind. Again, long distance wind transport of sand from Kordofan did not occur.

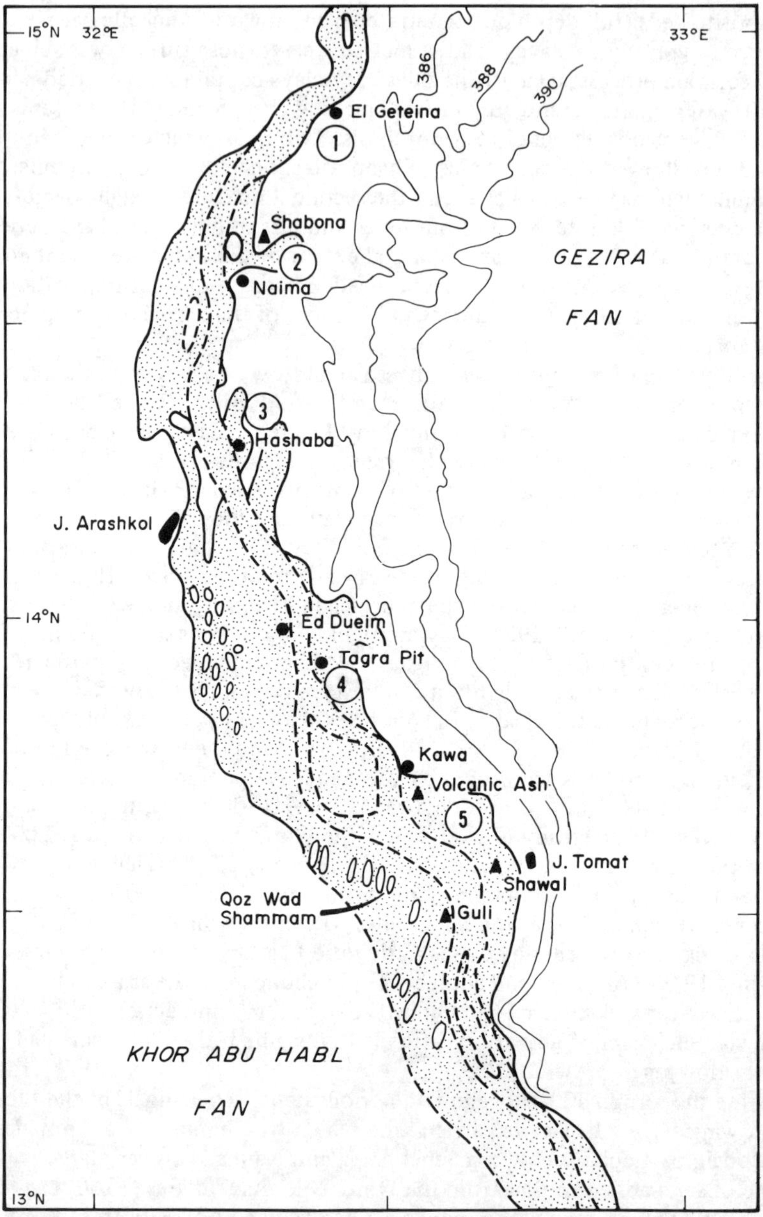

Figure 9.11. The maximum extent of terminal Pleistocene to early Holocene flooding of the White Nile trough in the period c. 11 000 to 12 000 BP is shown by the shaded area. The unflooded islands are sand dunes. On the eastern side of the flooded area, five embayments occur where contour lines on the Gezira curve away from the White Nile. The maximum present extent of flooding by the Jebel Aulia Dam is approximately indicated by the dashed lines within the shaded area.

The arid period of the late Pleistocene was one of aggradation within the White Nile channel as well as on the Gezira. As with the Blue Nile palaeo-channels, the White Nile was not incised and it probably changed course frequently as a result of aggradation and the movement of wind-blown dunes. Seasonal floods would cause ponding, force new channels through the sand-choked trough, and supply water to evaporation-basins which existed in deserted channels and on lateral floodplains.

9.3.2 *Terminal Pleistocene and Early Holocene flooding by the White Nile*

The supply of water from Lake Victoria to the Sudan after about 12 000 years ago transformed environmental conditions along the White Nile at the same time as discharge from the Blue Nile increased greatly. The almost horizontal White Nile trough became inundated, particularly when the peak of the Blue Nile floods ponded the White Nile (Figure 11). Even in the low season of river flow, extensive areas would have remained permanently flooded and have been connected with the river because the river was receiving a substantial continuous discharge from the central African lakes. When this discharge first poured into the Sudan along the Asua River-Bahr el Jebel drainage line, the Sudd swamps would not have been as extensive, or possibly were not even present. The regulating effect of the swamps on river flow downstream of the Sudd would have been absent, resulting in greater seasonal fluctuation than occurs today. Very high discharge by the White Nile in the absence of buffering by the swamps could have contributed to the exceptional floods along the main Nile at this time. As the Sudd swamps extended they reduced the seasonal fluctuation in peak- and low-flow of the White Nile, and stabilised the area of permanent flooding along the lower White Nile.

The new lacustrine conditions were rapidly established between 11 000 and 12 000 years ago. Lacustrine clays containing the shells of permanent-water mollusca were deposited over former soil surfaces, as shown in the profile from Esh Shawal (Figure 9). Although the seasonal fluctuations of the White Nile were much less than before, the flatness of the country flanking much of the river ensured that a wide strip of land was also seasonally-flooded to create 'toich' conditions similar to those existing today in southern Sudan.

In the early stages of its flood regime the White Nile may have forced extra channels between the choking sand accumulations in the Ed Dueim-El Geteina area. The sand-dunes in the river trough became islands or peninsulas (Figure 11), flanked by permanent water or seasonally-flooded toich. As described in chapter 7, the sand bodies became subject to erosion and accretion by wave action. At high water levels, a fetch of 10 to 15 km would not be unusual.

The existence of a narrow, shallow lake and swamps along the White Nile is well accepted (Tothill 1946; Williams 1966; Williams & Adamson 1973, 1976) and its age has been firmly bracketed by radiocarbon dating (SUA-68, I-1485, SUA-75, I-1486, SUA-298) within the interval 7 000 to 11 500 years ago. The maximum extent of flooding at the close of the Pleistocene and during the early Holocene is shown in Figure 11. On the eastern side of the White Nile are a series of prominent embayments in the Gezira plain. They

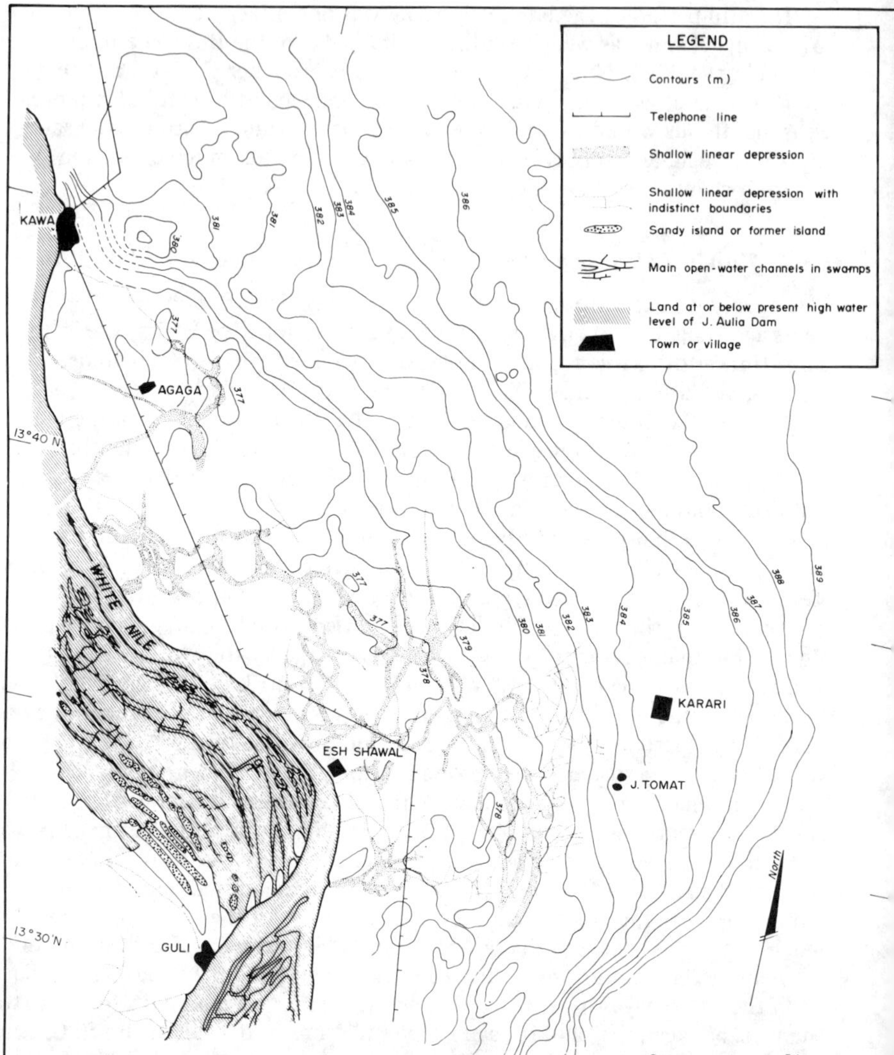

Figure 9.12. The Esh Shawal Embayment. The contours are generalised from the Roseires Soil Survey, White Nile E. Bank Rabak-Khartoum, Area 5A, Soils and Engineering Reconnaissance Plans, Ministry of Agriculture, Republic of the Sudan carried out by Hunting Technical Services and Sir M. Macdonald and Partners 1964. The locations of the shallow linear depressions, sandy islands and channels in the swamps were determined from aerial photographs. Because of the errors involved in superimposing these two types of information, and the errors involved in drawing contours on the low-relief floodplain, discrepancies are to be expected between the positions of contour lows and the depressions.

reflect curvature of the contours away from the White Nile. For example, at the northern end of the Esh Shawal embayment, the contour lines which mark the outer edge of the embayment converge on the White Nile at Kawa (Figure 12). The margins of the two northerly embayments are obscured by the sands which poured in along two major Blue Nile palaeochannels during the late Pleistocene (Figure 13). Because of the sand deposits, only limited areas in these two embayments were subjected to flooding from the White Nile in the Holocene. Further south, in the embayments which were not choked by sand, extensive clays and shell-deposits were laid down under permanent water in the early Holocene. In many cases, particularly in the Hashaba and Tagra embayments, the Holocene deposits form a thin skin over saline sediments that are probably inherited from the late Pleistocene intervals of highly seasonal flow in the White Nile.

The northern end of the White Nile trough, roughly between Jebel Aulia and El Geteina, is constricted between the sands derived from the Gezira palaeochannels and continuous outcrops of bedrock along the western bank. This constriction formed the northern end of the extensive early Holocene White Nile lake and swamps. During the last late Pleistocene arid period this part of the trough was probably choked with sand from the Blue Nile distributaries emerging from the Gezira fan as well as from the White Nile. When the discharge of the Blue and White Nile rose at the end of the Pleistocene and during the early Holocene, the pre-existing channels would have been inadequate. Restricted flow through this region would also have favoured ponding upstream.

Over a distance of up to about 15 km south of Jebel Aulia at least three palaeochannels are exposed in cross-section as a result of wave erosion on the eastern bank of the White Nile reservoir upstream of the Jebel Aulia dam. Three such sections are illustrated in Figure 14.

Section one is 2-3 km upstream of the dam, and is the most northerly of the three bank sections examined in the company of Drs Boushi and Abdel Salam, who first drew our attention to these sites. In common with sections two and three, the exposed channel-fill is up to 4 m thick, up to 150 m wide, and upwardly convex; in addition, the channel-fill in section one clearly occupies a depression in the underlying calcareous sandy clay. The base of the other two sections was not seen, and must be below the December 1972 White Nile reservoir level.

The upper 50 cm of section one consists of an admixture of sherds, bone fragments, rolled quartz pebbles and broken shells in a matrix of fine sand. Several unbroken pots were observed in the bank, together with large flecks of charcoal. The 195 ± 90 BP age for charcoal sample SUA-213 which came from 3 m above river level clearly indicates intrusion of much younger material into the southern edge of the channel-fill deposit. Intrusion may have been by scuffage and treadage, or via a faunal burrow. A bulk sample of shell fragments from a depth of 0.5 m (SUA-295) yielded an age of 5 480 ± 110 BP, indicating that the upper half-metre of the channel fill is of middle Holocene age, and that the bones and sherds are contemporary with Neolithic sites in the vicinity of Khartoum.

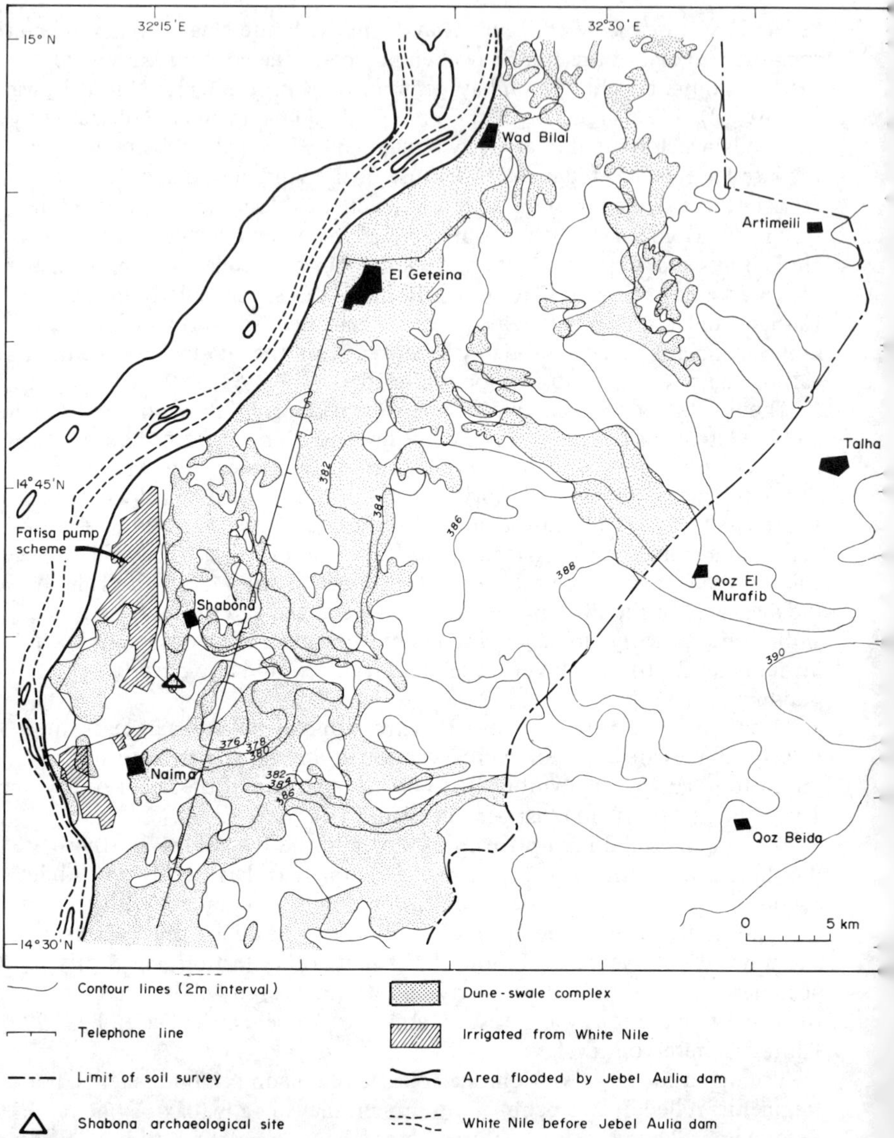

Figure 9.13. El Geteina and Naima-Shabona Embayments, showing the approximate distribution of the complex of sand dunes and clay-floored inter-dune depressions. For the source of information, refer to the legend to Figure 9.12.

On its northern flank the channel-fill overlies a calcareous sandy clay containing 30-50 % of soft white *in situ* carbonate aggregates and 5-10 % of hard black transported carbonate nodules. A date on the soft carbonate immediately beneath the channel-fill (SUA-296) of 10 320 ± 150 BP indicates that the sandy clay beneath the channel was already in existence by earliest Holocene times. Whether carbonate precipitation took place before, during or after the active life of the palaeochannel we cannot say. Certainly, the slight upward-coarsening evident in the channel deposits, the absence of cut-and-fill structures within the channel, and the fairly homogeneous nature of the fill suggests that it was laid down over quite a short interval. We conclude that the channel was active at some stage(s) between about 10 000 and 5 500 BP, and was virtually defunct by 5 500 BP.

The second section shown in Figure 14 is of interest in that the hard carbonate gravels within the fill are clearly allochthonous, and are disposed in horizontal bands within a matrix of fine quartz sand. Sample SUA-383 collected from a depth of 1.8 m (1.2 m above river level) gave a date of 19 450 ± 340 BP, in contrast to I-9291 (>40 000 BP) which was beyond the range of radiocarbon dating, although stratigraphically somewhat younger. This latter sample came from a depth of 1.2 m (1.8 m above water level), or from the same level as the soft *in situ* carbonate sample I-9292, which was dated at 12 640 ± 180 BP, indicating deposition of the channel sands and gravels before the start of the Holocene.

Bone, shell and pottery fragments were scattered through the upper 0.3-1.0 m of the section, embedded in a matrix of fine sand. Two hundred metres further south was what seemed to be a separate channel-fill, which was also replete with carbonate concretions and shell fragments.

Section three consists of a very rich surface concentration up to 30 cm thick of pottery, grindstones, polished stone axes and adzes, pendants, quartz flakes, bones, teeth, and charcoal within a matrix of fine sand, overlying 3 m of fine alluvial sands and clayey sands. The site is backed by sand-ridges some 200-400 m to the east, and is almost due east from Jebel Bereema, which is now an island in the White Nile reservoir, but was probably not an island during the Neolithic or Jebel Moyan occupation of this area.

The conclusion we draw from the channel-fill stratigraphy is that the channels were transporting fine sands and a bed-load of carbonate gravels of mixed age during the closing stages of the Pleistocene, and had become aggraded to their present form by Neolithic times or earlier. The prehistoric settlement site at section three would undoubtedly repay detailed examination, following a careful survey of prehistoric sites bordering the present White Nile between Jebel Aulia and El Geteina.

As the Blue Nile incised during the Holocene the level of flooding by the White Nile also fell. The effect was to reduce the lateral extent of seasonal and permanent inundation. The present land surface in the areas affected by Holocene flooding shows very clear signs of the flooding and of its retreat. Aerial and satellite photographs show a reticulum of shallow elongated depressions forming a geometrical pattern of darker lines across the Holocene floodplain (Figure 12). The occurrence of this pattern indicates the presence

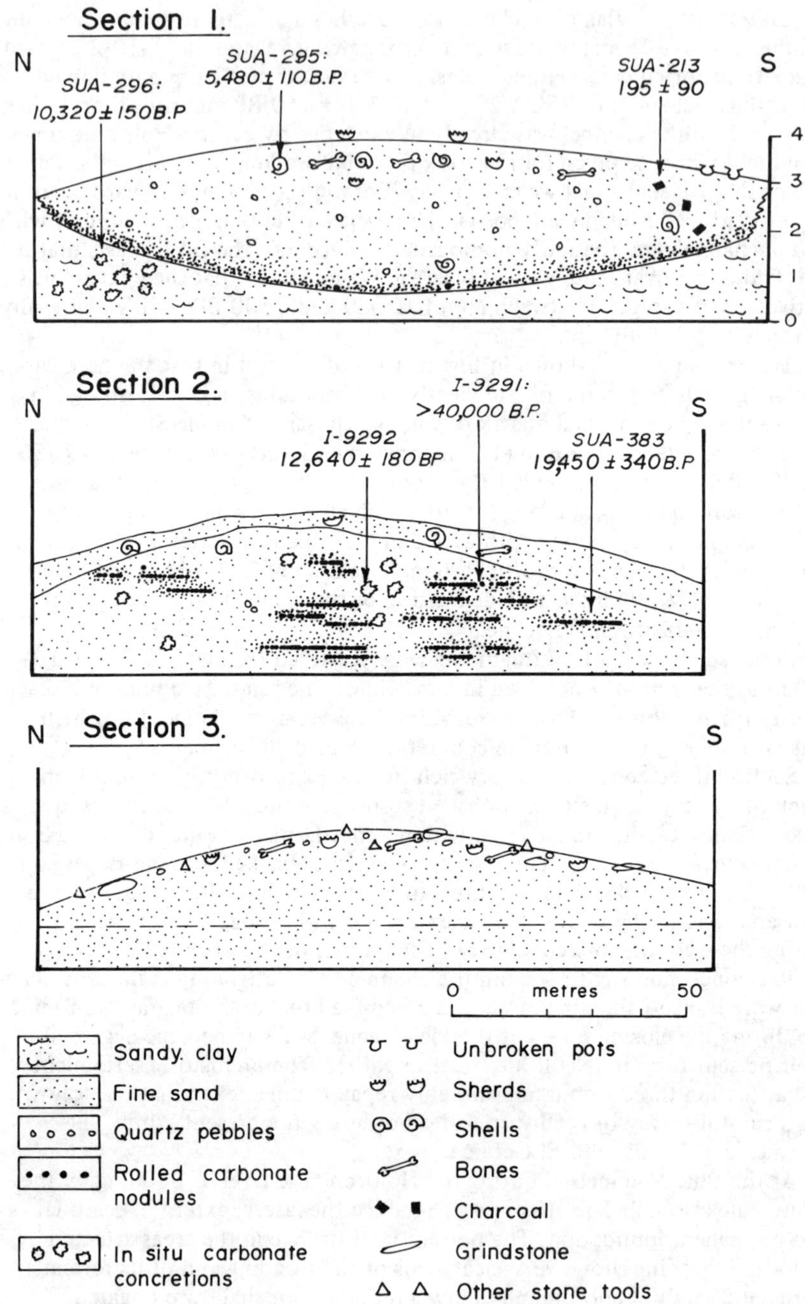

Figure 9.14. Cross-sections of palaeochannels exposed in the east bank of the White Nile south of Jebel Aulia Dam.

of former permanent channels and, in some embayments, high levels of sub-surface salinity. A present day analogue is found in the White Nile swamps on the western side of the White Nile opposite Esh Shawal and north of Guli, and at the northern tip of Aba Island in the White Nile opposite Guli (Figure 12). Future drainage of this area would undoubtedly result in a clay plain with a prominent pattern of linear depression corresponding to the deeper channels in the formerly flooded area. The shallower areas correspond to relict point bars, some of which emerge today as islands or near-islands in the swamps. The Guli prehistoric midden, dated at 5 500 ± 100 BP (SUA-211) using *Pila* shell, contained a barbed bone point (Adamson, Clark & Williams 1974) and was located on one such island (Figure 12). The deeper areas, which become the linear depressions after drainage, are the swales between the point bars and the inter-connected channels through which the rising and falling flood waters move. On the eastern side of the White Nile in the Shawal Embayment some of these linear depressions are aligned in accord with river flow along the curving eastern margin of the embayment. This is particularly clear 3 to 5 km to the west of Jebel Tomat (Figure 12).

During the progressive slow incision of the White Nile, conditions on the flood plain changed gradually from permanent flooding to seasonal flooding of the general surface with permanent water in the shallow channels. At a later stage of incision, only the channels received water at the peak flow of the White Nile and ultimately even this input was cut off as further incision occurred. In a similar but less spectacular fashion to the Gezira palaeochannels from the Blue Nile, these depressions were beheaded by incision. The upper two metres of the stratigraphic column illustrate this sequence clearly (Figure 9). Today the shallow depressions on the Holocene flood-plain collect local rain or irrigation water, but when crossing the clay plain it is extremely difficult to detect their presence. They have not been described previously in this region because of the difficulty of detecting them on the ground. Where they impinge on the present banks of the White Nile they are more obvious because they flood for short distances from the river and form 'maya'as'. Networks of these depressions which connect with the present eastern bank of the White Nile to form maya'as are shown in Figure 12.

9.3.3 *Holocene environments at Jebel Tomat and Shabona*

Jebel Tomat is a typical Sudanese inselberg in that it rises abruptly from the surrounding clay plains, emerging as a solitary granitic island from a sea of alluvial clay. 'Tomat' or 'el Tuweimat' means 'the twins', for the jebel is in fact two rocky hills some 30-40 m high separated by a broad, level col. The col is floored by stony, sandy and clayey detritus washed down from the two adjacent rocky hills, and has been used intermittently by man for several thousand years (Clark & Stemler 1975; Williams et al. 1975).

The jebel is located on the margin of a low but distinct scarp separating two broad clay benches (Figures 15 and 16). The lower, western bench rises eastwards from about 378 m to about 382 m over a distance of roughly 8 km between Esh Shawal and Jebel Tomat, illustrated in Figure 16. The higher,

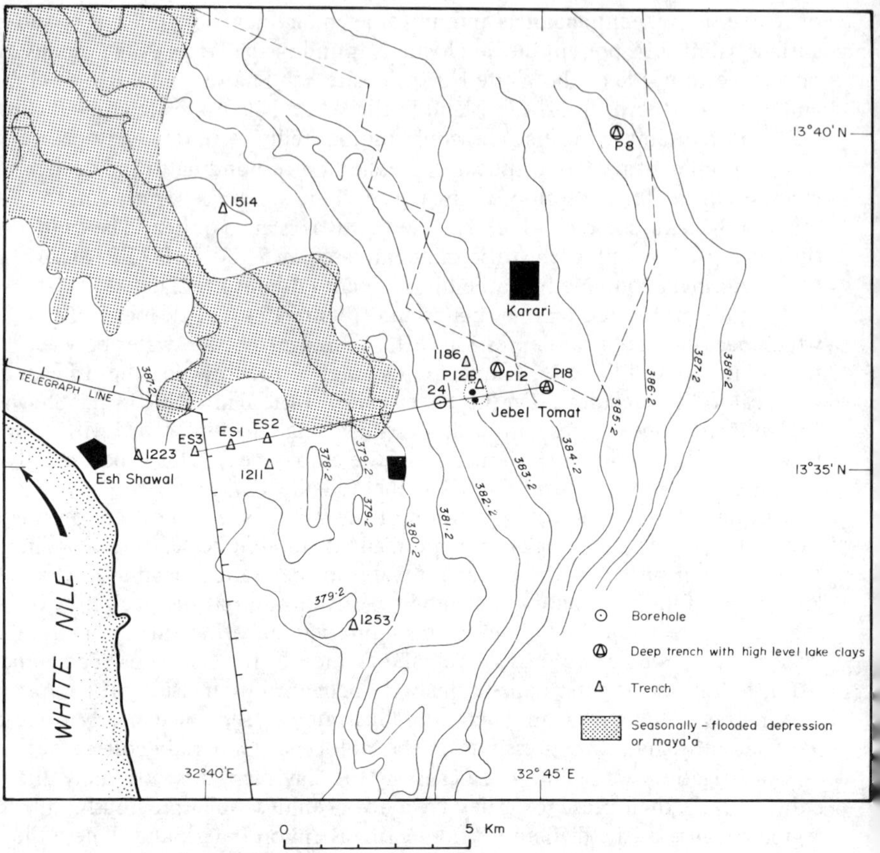

Figure 9.15. Location of Jebel Tomat-Esh Shawal stratigraphic sections.

eastern bench rises eastwards from about 384 m immediately east of Jebel Tomat to c. 387-388 m some 7 km to the east-south-east, where the land again rises steeply. These two benches become more prominent further south, between Rabak and Malakal, where they were surveyed by Gunn and colleagues during 1951-53 (Gibb 1954; chapter 6 of this volume). Berry (1960-61) referred to these two clay benches as the '382 metre terrace' and the '386 metre terrace', and both he and Whiteman appear to regard them as terminal Pleistocene erosional features (Berry & Whiteman 1968; Whiteman 1971:114-117), a view we wish to qualify.

In order to assess when and whether the landscape around Jebel Tomat had been eroded (or built up) by the White Nile, we described and sampled a large number of trenches and boreholes (some of which are shown on Figure 16) running from an elevation of 379 m two kilometres east of Esh Shawal to an elevation of c. 384.5 m two kilometres east of J. Tomat. The deepest trench section was 3.5 m, deepened to over 5 m by augering into the floor of the trench. The essential information obtained from this exercise may be interpreted as follows:

1. During the late Pleistocene, microcrystalline dolomite and high Mg-calcite accumulated in what was evidently a desiccating White Nile lake, the extent of which remains unknown. The radiocarbon ages we have so far obtained for these evaporite deposits range from >40 000 (SUA-382) through 36 800$^{+3\,100}_{-2\,200}$ (SUA-381) to 25 600 ± 700 BP (SUA-837).

2. The final phase of evaporite deposition at site ES3 was preceded by a brief interval of sedimentation of clayey sands, and was followed by the accumulation of further clayey sands, this time more gravelly.

3. Following deposition of the gravelly clayey sands, up to a metre of olive-grey fine sands was laid down under reducing conditions, presumably as part of the bed-load of the late Pleistocene White Nile. The fine particle size, uniform quartz mineralogy, and high degree of sorting of these sands suggest that they may well be reworked wind-blown sands derived from the pre-Holocene dunes that are so common along the lower reaches of the White Nile.

4. Deposition of the fine grey sands gave way to a brief interval during which were deposited clayey sands and abundant rolled carbonate nodules. This interval marks a transition in White Nile channel regime from a bed-load channel to a suspended-load channel, which we cannot date precisely, but know to be somewhat older than about 11 300 BP.

5. Assuming that the Pleistocene-Holocene boundary is at 11 000-10 000 BP, the interval of time immediately preceding the Holocene represents a period during which the White Nile was permanently flooded up to an elevation of at least 378 m in the vicinity of Esh Shawal. By about 11 300 BP water level had probably fallen to 378 m, for the concentration of shells of both juvenile and adult freshwater mollusca at this level denotes death from exposure along an emergent littoral zone. The 11 300 BP strandline east of Esh Shawal represents a still-stand during a progressive fall in White Nile level rather than a pause during a transgression, since there are no olive-green fluvio-lacustrine clays above any of the dated or sampled shelly layers.

Berry & Whiteman (1968) have expressed doubts about the precise geomorphic significance of the 11 300 BP shell age (I-1486) reported by Williams in 1966, on the grounds that the molluscs could have died on the bed of the river rather than along the shore of the much expanded end-Pleistocene White Nile.

We believe that the field evidence now available to us indicates that a concentration of dead shells at an elevation of c. 378 m at two sites several kilometres apart almost certainly took place simultaneously: SUA-75 (11 250 ± 220 BP) and I-1486 (11 300 ± 400 BP) are not significantly different in age. Since then, the perennial White Nile has not transgressed any of the stratigraphic sites shown on Figure 16. The '382 metre terrace' is strictly speaking

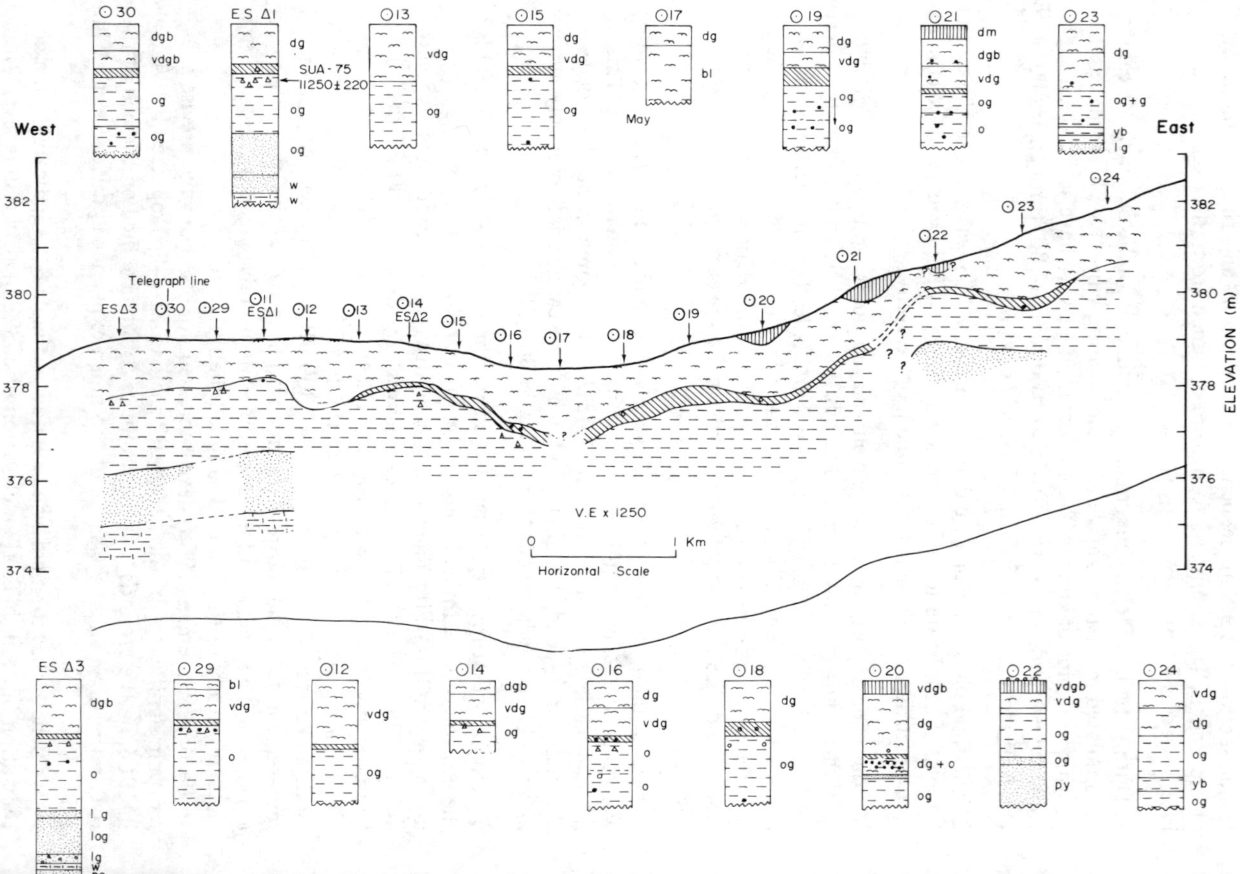
West
East
ELEVATION (m)
382
380
378
376
374
Telegraph line
SUA-75
11250± 220
V.E x 1250
0
1 Km
Horizontal Scale
ES Δ3
ES Δ1
ES Δ2
May

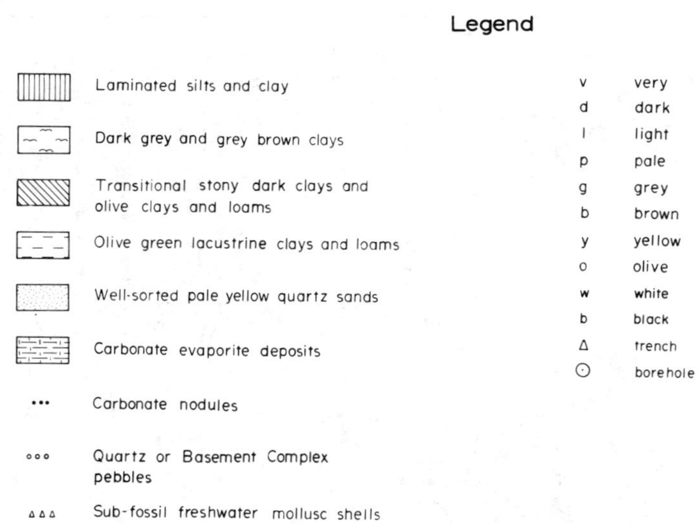

Figure 9.16. Detailed stratigraphic sections between Jebel Tomat and Esh Shawal.

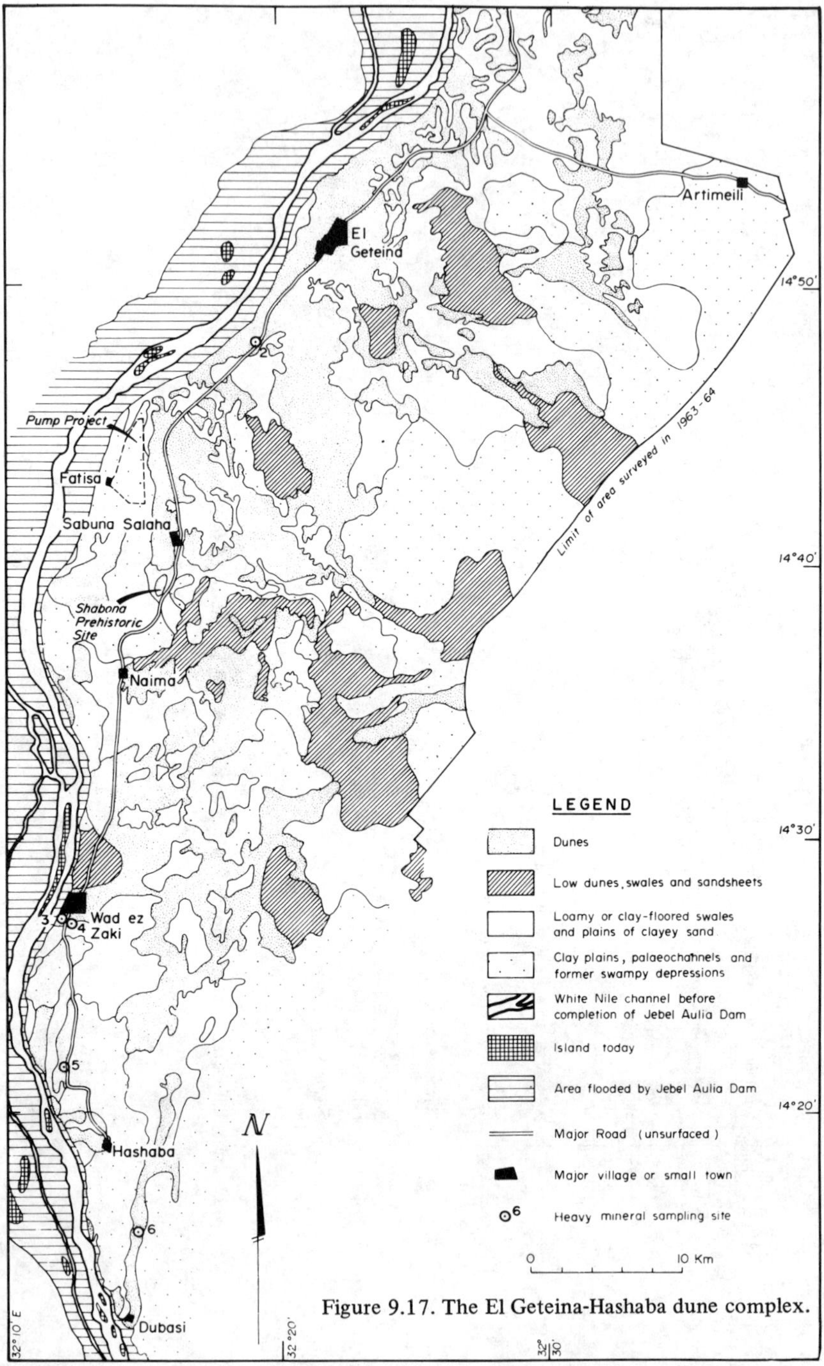

Figure 9.17. The El Geteina-Hashaba dune complex.

a depositional feature, not an erosional one, and is underlain by over a metre of dark cracking clay, which we will now consider.

6. Shortly after 11 300-11 250 BP, regression of the White Nile exposed the olive-green fluvio-lacustrine clays to subaerial weathering and erosion. Occasional fragments of vein quartz and of other Basement Complex rocks immediately overlying the green clays mark the unconformable contact between green and black clay. This regression probably coincides with the five centuries or so during which the level of Lake Victoria fell and overflow into the White Nile ceased towards 10 000 BP (Kendall 1969).

7. A renewed transgression soon after 10 000 BP brought a seasonal increment of dark clays to the country east of Esh Shawal. Earlier drainage channels entrenched into the green clay became infilled with black clay by these early Holocene White Nile floods. Borehole site 17 on Figure 16 is a case in point.

About 30 km downstream, near Tagra village, unbroken amphibious and freshwater mollusc shells near the base of the dark clays gave ages of 8 700 ± 350 BP (I-1485, depth 1.6-1.9 m) and of 8 130 ± 225 BP (SUA-68, depth 1.2-1.4 m). At a depth of 1.3 m Adamson also recovered two fragments of a barbed bone point, fragments of small bones of mammal, and fish bones, including cat-fish spines (Adamson et al. 1974). The surface elevation of this trench is 378.4 ± 0.2 m which is 3 m above the local mean flood level (375.4 m) and 3.2 m above the mean low water level (372.2 m) of the White Nile (375.4 m) at the nearby Ed Dueim gauge before completion of the Jebel Aulia dam. We conclude that the early Holocene White Nile attained a flood level at least 3 m above its modern unregulated flood level at Tagra towards 8 000-8 500 BP.

The 11 300-11 250 BP shells near Esh Shawal were deposited when the White Nile flood level was about 378 m, which is 5 m above the mean minimum and 2.5 m above the mean maximum river level at Esh Shawal before the dam was built. Owing to the cracking nature of the upper dark clay at Tagra, Esh Shawal and elsewhere along the White Nile, no obvious sedimentary structures or planes of stratification are evident within these early Holocene deposits, so that additional information on the middle to late Holocene history of this locality must be sought elsewhere.

8. A shallow trench dug at the eastern edge of the col at Jebel Tomat in January 1972 revealed the unconformable contact between the dark cracking clay and the overlying artifact-bearing colluvium. A sample of unbroken *Limicolaria flammata* shells from a depth of 60-80 cm at the base of the colluvium (SUA-67) yielded a minimum age limit for the dark clay and a maximum age for the prehistoric occupation site of 4 500 ± 130 BP. We conclude that much of the dark clay between Esh Shawal and Jebel Tomat was laid down between about 10 000 and 4 500 BP. Seasonal flooding by the White Nile, and trapping of clay-sized particles in the swampy maya'as and drainage channels bordering the river resulted in the accumulation of over 150 cm of dark cracking clay during an interval of perhaps 5 000 years. This mean rate of sedimentation (0.3 $mm.yr^{-1}$) was almost certainly far slower than the rate of accumulation

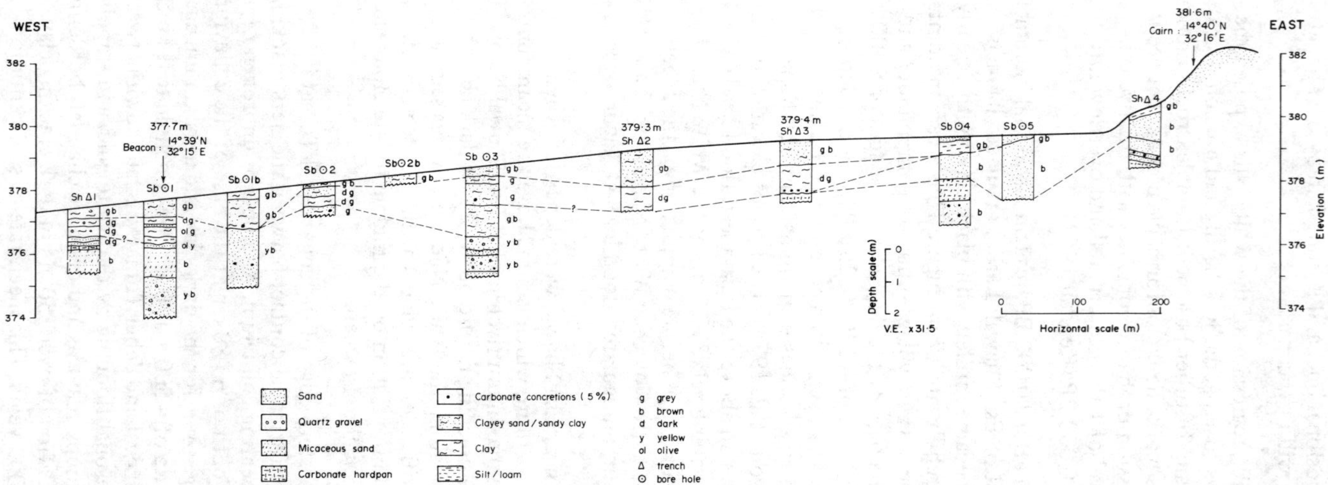

Figure 9.18. Transect east of Shabona prehistoric site.

of the shell-bearing fluvio-lacustrine green clay, which very probably began with overflow from Lake Victoria towards 12 500 BP (Kendall 1969), and ended towards 11 300 BP, giving a mean sedimentation rate of roughly 1 mm. yr^{-1}. The differences in depositional milieu, one a lake floor, the other a seasonal swamp, no doubt accounts for the colour and structural differences between the two clays.

9. The late Holocene prehistoric occupation of Jebel Tomat will be considered in detail in another volume, so that only the essential points need to be retained here. The base of the deposit containing the 4 500 BP shells also contained fossil bones and quartz flakes, but major occupation of the site was not effective until about 2 000 years ago (1 930 ± 60 BP to 1 600 ± 80 BP, Clark & Stemler 1975). The pastoral nomads who frequented this site raised cattle and sheep/goats, cultivated sorghum, ate fish, and collected *Pila wernei* shells for meat or bait. Since the White Nile is now a day's walk from the site, swamps full of fish and mollusca must have existed close to the jebel at that time. The reticulate network of defunct drainage channels shown on Figure 12 was probably at least seasonally active at that time. The existence of swamps near to the jebel in what is now a dry clay plain suggests that the local rainfall was higher between about 1 600 and 1 900 BP and/or the White Nile floods were more extensive at that time.

We will now turn to Shabona, which is c. 100 km downstream of Esh Shawal, and one degree of latitude further north. The general physiographic setting of the prehistoric site is shown in Figures 13 and 17, the detailed cross-sections in Figures 18 and 19, and the generalised stratigraphy in Figure 20.

Deposition of the fluviatile quartz gravels, coarse and medium sands and occasional bands of rolled carbonate nodules was accompanied or followed by aeolian reworking of some of the point-bar sands. The main sand-ridge on which the Shabona prehistoric site is situated has the appearance of a wave-trimmed foredune which probably originated as a channel-bar which then became reworked by wind-action.

Occupation of the site goes back to at least 7 050 ± 120 BP (SUA-298), which is the age of the *Pila* shells collected from the upper 20-30 cm of one of the shallow prehistoric shell-middens on the summit of the sand-ridge. *Pila* shells occur in the dark clays and loams both east and west of the site, and deposition of the basal clays east of the sand-ridge may have been under way by about 7 000 BP, if not sooner. Swamp conditions persisted at least seasonally until somewhat after 2 700 ± 140 BP (SUA-299), which is the age obtained on *Pila* shells embedded in dark clay at a depth of 100-120 cm to the east of the main site (Figure 18). A further date of 3 030 ± 80 BP (SUA-70) from a sandy island opposite Tagra, some 50 km upstream, was also from *Pila* shells associated with broken pottery, as was the shell-date of 5 500 ± 100 BP (SUA-211) from the barbed bone-point site at Guli, opposite Esh Shawal.

With radiocarbon dates of c. 8 500-8 000; 7 000; 5 500; 4 500; 3 000; and 2 000-1 500 BP from shells and carbonised samples associated with prehistoric settlement sites along the lower White Nile, it seems that Holocene occupa-

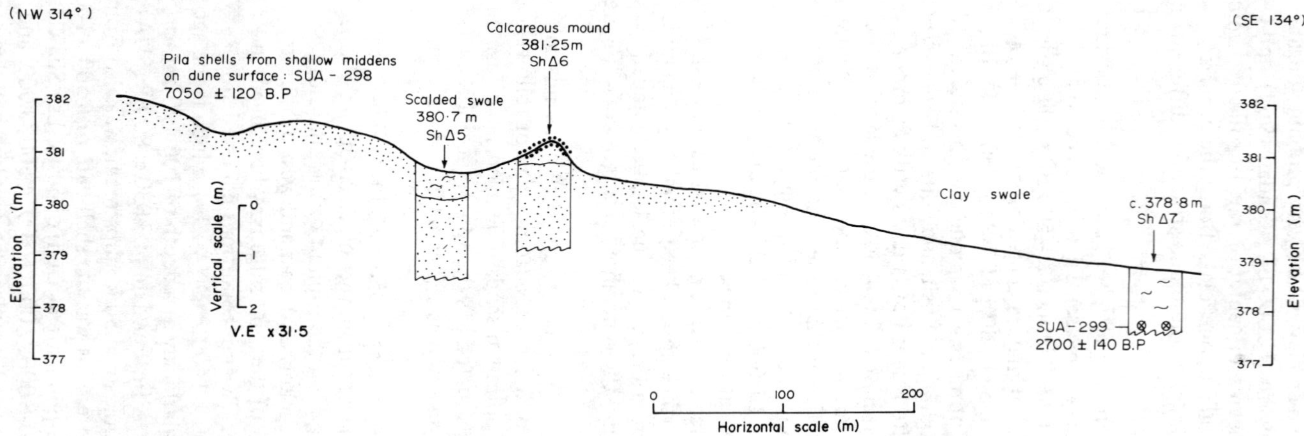

Figure 9.19. Stratigraphic sections west of Shabona prehistoric site.

tion of this area has been virtually continuous since the final recession of the pre-Holocene White Nile lake towards the onset of the Holocene some 10 000 to 11 000 years ago. Certainly, there have been intervals when local rainfall was higher during the Holocene, as the existence of small lakes 20 to 40 m above the White Nile flood level in the Wad Mansurab basin 20 km northwest of Jebel Aulia towards 8 500-7 000 BP clearly indicates (Williams et al. 1974). The presence of swamps at Shabona towards 2 700 BP, and at Jebel Tomat towards 1 900-1 600 BP, are very likely a result of somewhat wetter conditions at those times. Only a great deal more digging at appropriate sites in this region that is so lacking in good stratigraphic exposures will allow us to further refine the palaeogeographic reconstructions outlined above.

9.4 CONCLUSION

9.4.1 *Chronological summary of palaeogeography and palaeoenvironments*

A chronological summary of the main geographic events and environmental conditions in the Holocene and late Pleistocene follows. In broad terms, the terminal Pleistocene was locally arid with active dune formation, and both Niles were aggrading, unincised and very highly seasonal with lower discharges than at any time during the Holocene. At about 12 000 years ago the climate became wetter in the headwaters of both Niles; Lake Victoria overflowed to make the White Nile a steady flowing river feeding a developing lake and swamps at its northern end; discharge of the Blue Nile rose greatly and flooding and silt deposition spread across the Gezira via many distributaries. During the early Holocene, extensive flooding persisted on the Gezira and along the White Nile depositing the skin of fine Holocene sediment which we see today as alluvial clay. Downcutting by the Blue Nile, now consolidated into one or a few major channels, commenced and former distributaries and swamps on the Gezira fan were progressively beheaded. By middle Holocene times the incision of the Blue Nile was well advanced and its present channel was carrying the bulk of the discharge. The bed of the White Nile was lowered in accordance with that of the Blue and main Niles, slowly draining the shallow White Nile lake. Periods during the middle and early Holocene were locally wetter than they are at present. Incision by the rivers continued during the upper Holocene, leading to the final conversion of the Gezira into a plain totally dependent for water on local rainfall.

9.4.2 *Towards a coherent history of the Late Pleistocene and Holocene Nile basin*

As pointed out at the beginning of this chapter, the behaviour of the Blue and White Nile rivers in central Sudan must be consistent with events both upstream in Ethiopia and Uganda, and downstream along the main Nile in Egypt and Nubia.

Because the Gezira and its flanking Blue and White Niles are centrally

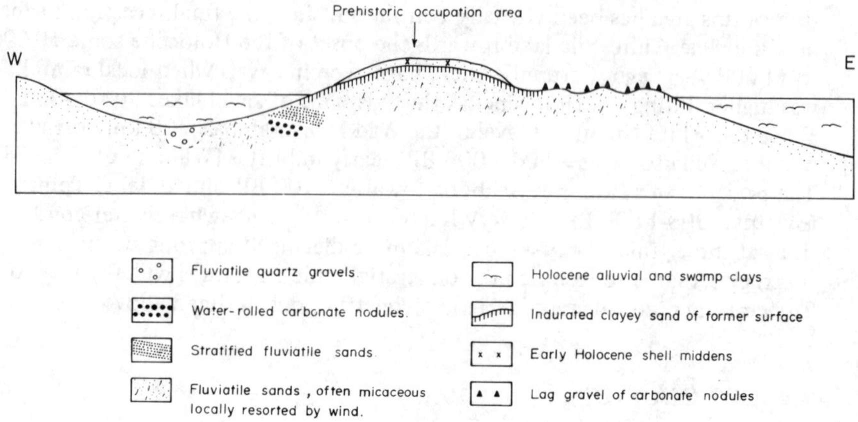

Figure 9.20. Generalised stratigraphic section, Shabona prehistoric site.

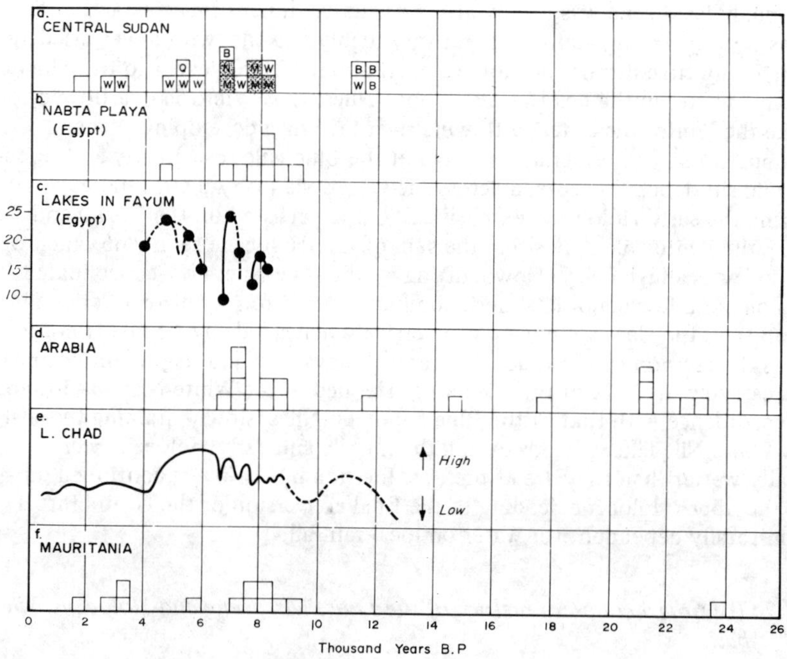

Figure 9.21. The distribution of radiocarbon dates, and fluctuations in lake levels, for selected localities during the late Pleistocene and Holocene. Each square represents one radiocarbon date. See also Chapter 10.

located within the Nile basin, evidence from this region is crucial to the integrated interpretation of the history of the whole Nile. Until the present work, the palaeogeography of the central part of the Nile basin was poorly known, as recognised by Butzer & Hansen (1968). By contrast, the palaeogeography around the Nile headwaters in East Africa, the Ethiopian and Afar Rift Valleys, and Chad, and along the lower Nile in Sudanese Nubia and southern Egypt is now well documented. The broad pan-African correlations in the sahel and savanna zones were considered in chapter 2, and it is appropriate here to examine briefly how our data on the history of the Nile in the central Sudan match those from the opposite ends of this immense river system.

Correlations between the behaviour of the Blue and White Niles in central Sudan and the well-documented behaviour of lakes in Kenya, Uganda, Ethiopia and Chad are generally good. The example already discussed, of the correspondence in time between the overflow of Lake Victoria and the sudden appearance of permanent lake and swamp conditions along the lower White Nile more than ten degrees of latitude to the north, demonstrates the excellent correlation between events in the southern headwaters and in the central Sudan.

The outstanding problem is to obtain a correlation with events along the lower main Nile in Nubia and Egypt. The problem has been well stated by Butzer et al. (1972) who pointed out that relatively high floods inferred for the main Nile in Nubia between 12 000 and 17 000 years ago find no counterparts in the East African record. This is the crux of the problem. During this period which was markedly more arid than today in the areas around the headwaters of the Nile Basin, the main Nile in Egypt and in Sudanese Nubia is said to have had a discharge and competence markedly greater than today (Butzer & Hansen 1968; De Heinzelin 1967, 1968). Greater discharge did not come from the local wadis at this time and it could only have come from the

---

Central Sudan. Dates on shells obtained by the authors in the vicinity of the past or present White Nile (W) or Blue Nile (B). The shaded squares represent dates from sites influenced only by local climatic conditions, near Wad Mansurab (M) on the western side of the White Nile north west of Jebel Aulia, and at En Nahud (N) in Kordofan. The En Nahud site was sampled by the late Dr W.H. Morton during joint field work in western Sudan early in 1976. The date Q from El Qoz south of Khartoum on the White Nile is the average of two dates obtained by Arkell.

Nabta Playa. Playa lake remote from the Nile in the Egyptian Sahara, dates from Wendorf, Schild et al. (1976).

Arabia. Local lakes in Arabia, dates from McClure (1976).

Mauritania. Sebkhas, also reflecting local climatic conditions (Street & Grove 1977).

Lakes in Fayum. The fluctuations in the level of water in the Fayum Depression in lower Egypt reflected changing behaviour of the Nile, as the depression was mainly fed by flood water from the Nile. Modified from Wendorf & Schild (1976:223).

Lake Chad. Relative lake levels, replotted from the curve presented by Maley (1977) who used the data of M. Servant (1973) and S. Servant (1977) for the period to 9 000 BP and including the probable recession around 6 600 BP. The curve is extended to 12 000 BP using the arid-humid curve for L. Chad presented by M. Servant (1973).

headwaters, where regional aridity dominated. Williams & Adamson (1974) also recognised this problem.

The problem has not yet been resolved. For example, Hassan (1976) describes the deposition of that part of the Gebel Silsila Formation (Butzer & Hansen 1968) which corresponds to the Sahaba unit (De Heinzelin 1968) as coinciding 'with Nile floods of high amplitude and, seemingly, an increase in the contribution from the non-Ethiopian tributaries south of the Sahara'. Yet this period from c. 12 000 to 17 000 years ago was notably arid in regions where these non-Ethiopian (i.e. White Nile) tributaries arise. The dilemma extends into the Holocene period. Smith (1976), referring to the period from about 12 000 to 5 000 years ago, concludes that 'the hydrological budget of the Nile itself now tended to be reduced as a result of climatic changes in East Africa'. This is the reverse of the trend to be expected from the East African data (chapter 2) and from our own data from the central Sudan.

Figure 21 presents data from the approximate latitude of the central Sudan and from further north. The distribution of radiocarbon dates during part of the late Pleistocene and the Holocene is shown for central Sudan, Egypt west of the Nile, Arabia and, as a remote comparison, Mauritania. The levels of Lake Chad and the Fayum Lakes in Egypt are also shown. These data show the remarkable general correspondence which applies across Africa and into Arabia and which is worth detailed analysis. For the present purpose, we will note that there is little sign of prolonged localised (Arabia, Nabta Playa, local depressions in central Sudan) or regional (White and Blue Nile in central Sudan, Lake Chad, Fayum Lakes in Egypt fed by Nile overflow) abundance of water in the period preceding 12 000 years ago. These results imply strongly that conditions were on average much drier from 12 000 to 17 000 years ago throughout the Nile Basin than in the early Holocene.

There are, of course, extremely important fluctuations which are concealed by the above generalisation. For example, the dates from Arabia suggest only episodic moist intervals until a substantial improvement in local climate between about 6 000 and 9 000 years ago. Remarkably, the interval from about 6 500 to about 9 000 years ago was capable of supporting local bodies of water in central Sudan, the Egyptian Sahara (Nabta Playa) and Arabia. Another example of probable significance is that the interval between about 9 500 and 10 500 years ago was probably rather dry locally and in a wide regional sense. Such a finding has important implications for the palaeogeographic picture we have built up for the Gezira and for the Blue and White Niles in central Sudan. Obviously a higher resolution of the palaeoclimatic and palaeoenvironmental reconstruction is possible.

Overall, the data in Figure 21, elsewhere in this chapter, and in chapter 2, strongly support the contention that correlation is now possible throughout the whole Nile Basin, as foreseen by Butzer & Hansen (1968, p.7) when they said 'It may ultimately be possible to correlate palaeoclimatic events as well as prehistoric cultures along the length of the Nile corridor'. It is likely, in our view, that the apparent lack of congruence between the stated behaviour of the main Nile in Egypt and the evidence from the rest of the Basin is caused by the particular interpretation of fluvial activity chosen. The solution to the dilemma in the period 12 000 to 17 000 years ago may lie in the direction of interpreting the 'vigorous . . . flood regime . . . including more extensive and probably longer lasting inundations, rapid shifts of a stream channel, a tendency to stream braiding, and a higher groundwater table' (Butzer & Hansen 1968:182) as the result of episodic high velocity floods from a very highly seasonal, unincised, aggrading river fed from headwaters in Ethiopia and East Africa at a time when regional climate was more arid than at present. The absence of a steady water supply from the White Nile during the annual low season of flow prior to 12 000 years ago must also have had a dramatic effect on the fluvial regime of the main Nile. Certainly, to interpret the fluvial activity of the main Nile in the interval 12 000 to 17 000 years ago as due to an increase in the mean annual discharge is not consistent with the local and pan-African palaeoclimatic evidence.

This is not the place for a full analysis of the fluvial history of the whole Nile in the late Pleistocene and Holocene. However the information now available from the central Sudan makes it abundantly clear that such an analysis is now possible. Evidence from the full 6 670 km of the Nile can now be integrated to produce a coherent picture of its late Quaternary history. The Gezira is as central to the interpretation of the palaeogeography of the Nile, as it is to the Basin itself. This proposition has been examined by Adamson et al. (1980) and is taken up in Chapter 10.

Table 9.1. Radiocarbon dates for central Sudan obtained by the authors.

Laboratory number	Location	Date year BP	Material	Remarks
I-9712	Khartoum-Wad Medani road, 500 m NNE of kilo 175. c. 14°29′25″N 33°27′30″E	23 780 ± 600	Carbonate	Pit I. Moderately soft nodules. In clay, 15 cm below base of volcanic ash deposit, in western face of pit.
SUA-605	Khartoum-Wad Medani road, 30 m W of road, near kilo 163. c. 14°35′N 33°26′E	24 920 ± 480	Carbonate	Pit III. Thin hard carbonate bands horizontally oriented within the volcanic ash deposit at 40 cm below surface
SUA-214	Gezira Agricultural Research Station, Agricultural Research Corporation, Wad Medani. c.14°24′N 33°30′E	11 975 ± 260	Shell	*Cleopatra* and *Lanistes* from pit dug for foundations of the Soil Survey Store building at depth of 160 to 170 cm in typical Gezira clay soil.

Table 9.1. (continued)

Laboratory number	Location	Date year BP	Material	Remarks
SUA-386	As for SUA-214	27 370 ± 800	Car-bonate	Small, hard, grey-black nodules c. 2 to 5 mm in diameter. From same pit and depth as SUA-214.
I-9289	Blue Nile, west bank, opposite Umm Dom Island c. 15°32′40″N 32°36′30″E	>40 000	Car-bonate	Hard large concentrations, in massive cemented calcrete, sampled 3 m below top of steep bank, exposed to flood water.
I-9290	As for I-9289	>40 000	Car-bonate	Semi-soft, from massive cemented calcrete. Sampled at depth of 1 m in pit, c. 40 m back from edge of bank which yielded I-9289.
I-9287	Blue Nile, west bank, near Manshia, c. 8 km upstream from Khartoum. c. 15°35′N 32°35′E	11 560 ± 160	Car-bonate	Porous, small lumps sampled from steep bank, c. 6 m above low water level.
I-9288	As for I-9287	7 830 ± 120	Car-bonate	Porous lumps sampled from same bank as I-9287. c. 4.5 m above low water level.
SUA-544	On Sennar-Singa road c. 30 km south of Sennar. c. 13°18′N 33°40′E	11 330 ± 150	Shell	Nile oyster, N bank of palaeochannel, at beacon and adjacent to railway and telegraph line. Shell fresh and pearly in appearance, Nile oyster 50 cm below surface, exposed in roadside.
SUA-604	Atra sand pit in kerrib, W bank of Blue Nile, opposite confluence of Blue Nile and Rahad. 14°28′N 33°29′30″E	11 800 ± 195	Shell	Date on *Cleopatra, Lanistes* and *Pila.* In poorly sorted current-bedded fluvial sands and gravels of palaeochannel of Blue Nile system.
SUA-215a	Wad es Said. 1-2 km SW of village in non-irrigated area. 14°40′N 33°13′E	6 890 ± 115	Shell	*Pila in situ* in fluvial sands showing current bedding.

Table 9.1. (continued)

Laboratory number	Location	Date year BP	Material	Remarks
SUA-213	East bank White Nile, c. 2-3 km S of J.Aulia Dam (Section 1).	195 ± 90	Charcoal	Charcoal in sandy fill of palaeochannel. Charcoal much younger than deposit, possibly in fill of faunal burrow.
SUA-295	As for SUA-213	5 480 ± 110	Shell	*Limicolaria* and *Pila* shell, 50 cm below surface of palaeochannel, associated with occupational debris, in fine sand of fill of palaeochannel.
SUA-296	As for SUA-213	10 320 ± 150	Carbonate	Soft white *in situ* carbonate partly composed of micro-crystalline calcite, located in calcareous sandy clay immediately beneath the channel fill sediments (SUA-295).
I-9292	East bank White Nile, c. 6-8 km S of J.Aulia Dam (Section 2).	12 640 ± 180	Carbonate	Soft white *in situ* carbonate, 120 cm below surface of palaeochannel, in deposit of horizontally stratified fine sands and hard carbonate gravel. Overlain by 0.3 to 1.0 m of fine sand containing occupational debris.
I-9291	As for I-9292	>40 000	Carbonate	Hard, rolled carbonate, nodules, horizontally bedded, 120 cm below surface, same layer as I-9292.
SUA-383	As for I-9292	19 450 ± 340	Carbonate	Soft white carbonate aggregates, irregular shape, partly composed of micro-crystalline calcite, 180 cm below surface, i.e. 60 cm below I-9292 and I-9291.
SUA-298	Shabona dune, archaeological site (Figure 13)	7 050 ± 120	Shell	Large, whole shells of *Pila* from shell-filled pit at 'Early Khartoum' culture site 20-30 cm below surface of sand dune.
SUA-299	Shabona, c. 0.5 km east of archaeological site	2 700 ± 140	Shell	*Pila* shell fragments, 100-120 cm below surface of clay plain.
I-1485	Near Tagra, about 2-3 km SSW of vil-	8 700 ± 350	Shell	*Pila* at a depth of 1.6-1.9 m below surface, near base

Table 9.1. (continued)

Laboratory number	Location	Date year BP	Material	Remarks
	lage and 0.5 km E of White Nile. Pit location 230 m on bearing of 42° from steel beacon marking 13°56′N 32°22′E.			of dark clay. During a later excavation, an extension of this pit yielded fish bones and fragments of barbed bone point at depth of 1.3 m.
SUA-68	As for I-1485 (same pit)	8 130 ± 225	Shell	*Pila* and *Lanistes* in extension of same pit as I-1485, which yielded fish bones and fragments of barbed bone points. Depth 1.2-1.4 m.
SUA-69a	As for I-1485	33 300 ± 1 150	Carbonate	Hard dolomite nodules, depth of 1.5-1.6 m in calcareous layer of fine sand and silt 10 cm thick, from same pit as I-1485 and SUA-68.
SUA-69b	As for I-1485	10 460 ± 160	Carbonate	Soft calcite concentrations. From same layer as SUA-69a.
SUA-70	Island in White Nile opposite Tagra village. c. 13°56′N 32°21′E	3 030 ± 80	Shell	*Pila, Lanistes, Etheria* mixed with broken sponge pottery beneath sand at present level of dam. Low sandy island c. 400 m W of eastern shore.
SUA-67	Jebel Tomat c. 13°36′N 32°44′30″E	4 500 ± 130	Shell	*Limicolaria,* at depth of 60-80 cm at contact between surface colluvium containing occupational debris and lower dark clay. On eastern edge of pediment between the two jebels.
I-1486	Esh Shawal, c. 4.5 km E of village. c. 13°35′N 32°41′E	11 300 ± 400	Shell	*Cleopatra, Corbicula, Etheria,* in shell bed at depth of 1.45-1.70 m on surface of greenish clay and beneath dark cracking clay.
SUA-75	c. 3.5 km E of Esh Shawal and 1 km SE of I-1486 site. c. 13°35′20″N 32°40′E	11 250 ± 220	Shell	From depth of 120-140 cm in shell-bearing green clay just below overlying dark cracking clay, and c. 240 cm above level of calcite sample SUA-297.

Table 9.1. (continued)

Laboratory number	Location	Date year BP	Material	Remarks
SUA-837	As for SUA-75	25 600 ± 700	Car-bonate	Microcrystalline calcite at depths of 4.1, 4.6 and 4.8 m respectively. Calcite separated from dolomite by acetic acid hydrolysis. Trench ES3.
SUA-381	As for SUA-75	$36\,800\ ^{+3\,100}_{-2\,200}$	Car-bonate	
SUA-382	As for SUA-75	>40 000	Car-bonate	
SUA-211	6.5 km NW of Guli village 13°33′N 32°34′E	5 500 ± 100	Shell	*Pila*, depth 0-0.35 m, on sandy point-bar, now an island (Gezira Birka Turab) rising c. 2 m above present swamps. Deposit a midden, occupational debris includes fish bone, sherds and fragments of barbed bone point.
SUA-385	3.5 km NW of Guli village 13°32′N 32°35′E	6 810 ± 110	Car-bonate	Small, dark grey-black smooth nodules up to 1 cm in diameter in upper 50 cm of point-bar bordering swamps (now partly reclaimed)
SUA-72	Wad Mansurab Basin, c. 13 km W of White Nile and 20 km NW of Jebel Aulia Dam 15°23′N 32°22′E	8 400 ± 160	Shell	Pit 1. *Biomphalaria, Cleopatra, Lanistes, Lymnaea, Melanoides* from a depth of 30-45 cm in calcareous silt. All the Wad Mansurab dates are from small basins dependant on local surface drainage.
SUA-74	As above	7 620 ± 130	Shell	Same assemblage as above, from a depth of 45-60 cm in Pit 1.
SUA-71	As above 15°22′30″N 32°21′30″E	7 870 ± 140	Shell	Pit 2a. *Cleopatra, Lanistes, Lymnaea, Pila, Biomphalaria, Melanoides* and occasional *Limicolaria* from a depth of 8-12 cm in a fine sandy loam.
SUA-73	As for SUA-71	6 990 ± 100	Shell	Pit 2b. *Bellamya, Lanistes, Limicolaria, Pila, Zootecus* from a depth of 12-20 cm.
SUA-606	En Nahud, Kordofan 12°44′N 28°26′E	6 800 ± 110	Shell	*Pila, Bulinus, Biomphalaria*, at depth of 2 m, from a well on the NE edge of a shallow basin adjacent to

Table 9.1. (continued)

Laboratory number	Location	Date year BP	Material	Remarks
				the market in Nahud. Shell layer is a cemented calcareous silt overlain and underlain by sand. This sample was collected by the late Dr W.H. Morton in February 1976.
SUA-545	Gezira Main Canal between Sennar and Wad Medani	112.4 ± 1.3 % Modern	Shell	Modern *Corbicula* recently dredged from canal bed during cleaning operations.
SUA-384	White Nile between Khartoum and J.Aulia Dam	145.2 ± 1.6 % Modern	Shell	Living mussels, shell only dated.
SUA-212	As above	144.5 ± 1.2 % Modern	Shell	Living *Corbicula*, shell only dated.
SUA-215b	White Nile c. 8 km S of J.Aulia Dam	146.7 ± 1.3 % Modern	Shell	Single shell of modern *Pila*, with operculum, dimensions 70 x 58 mm, weight 29 g.

All dates listed here are conventional radiocarbon ages corrected for isotopic fractionation, using an assumed group mean value of $\delta^{13}C = -5 \pm 2$ ‰ for shell and soil carbonate (Stuiver & Polach 1978). The changed correction values account for minor discrepancies between the dates in Table 9.1 and some previously published values.

9.5 REFERENCES

Adamson, D.A., F. Gasse, F.A. Street & M.A.J. Williams 1980, Late Quaternary history of the Nile. *Nature* 287: 50-55.

Adamson, D. & M.A.J. Williams 1973, Scientific results of the Macquarie University expedition 1972-73. 19th Ann. Rept. Hydrobiol. Res. Unit, Univ. Khartoum:17-20.

Arkell, A.J. 1949, *Early Khartoum.* Oxford University Press.

Arkell, A.J. 1953, *Shaheinab.* Oxford University Press.

Berger, W.H., R.F. Johnson & J.S. Killingley 1977, Unmixing of the deep-sea record and the deglacial meltwater spike. *Nature* 269:661-663.

Berry, L. 1960-61, The physical history and development of the White Nile. 8th Ann. Rept., Hydrobiol. Res. Unit, Univ. Khartoum:14-19.

Berry, L. & A.J. Whiteman 1968, The Nile in the Sudan. *Geogr. J.* 134:1-33.

Butzer, K.W. & C.L. Hansen 1968, *Desert and river in Nubia: geomorphology and prehistoric environments at the Aswan Reservoir.* Madison, University of Wisconsin Press.

Butzer, K.W., G.L. Isaac, J.L. Richardson & C. Washbourn-Kamau 1972, Radiocarbon dating of East African lake levels. *Science* 175:1069-1076.

Clark, J.D. & A. Stemler 1975, Early domesticated sorghum from Central Sudan. *Nature* 254:588-591.

Finck, A. 1961, Classification of the Gezira soil. *Soil Sci.* 92:263-267.
Gibb, Sir Alexander & Partners 1954, Estimation of irrigable areas in the Sudan 1951-53. Report to the Sudan Govt. London, Metcalfe & Cooper Ltd.
Hassan, F.A. 1976, Heavy minerals and the evolution of the modern Nile. *Quat. Res.* 6: 425-444.
Heinzelin, J. de 1967, Pleistocene sediments and events in Nubia. In W.W. Bishop & J.D. Clark (eds), *Background to evolution in Africa.* Chicago, University of Chicago Press: 313-328.
Heinzelin, J. de 1968, Geological history of the Nile Valley in Nubia. In F. Wendorf (ed.), *The prehistory of Nubia.* 1. Dallas, Texas, Southern Methodist University Press:19-52.
Hurst, H.E. & P. Phillips 1933, *The Nile Basin. III. Ten-day mean and monthly mean gauge readings of the Nile and its tributaries.* Physical Dept. Paper 29. Cairo, Government Press.
Kendall, R.L. 1969, An ecological history of the Lake Victoria basin. *Ecol. Monog.* 37: 25-52.
Maley, J. 1977, Palaeoclimates of Central Sahara during the early Holocene. *Nature* 269: 573-577.
McClure, H.A. 1976, Radiocarbon chronology of late Quaternary lakes in the Arabian desert. *Nature* 263:755-756.
Polach, H.A. & J. Golson 1966, Collection of specimens for radiocarbon dating and interpretation of results. Canberra, Austral. Inst. of Aboriginal Studies, Manual 2.
Servant, M. 1973, Séquences continentales et variations climatiques: Evolution du bassin du Tchad au Cénozoique supérieur. Unpublished D.Sc. thesis, Univ. Paris.
Smith, P.E.L. 1976, Stone age man on the Nile. *Sci. Amer.* 235:30-38.
Street, F. Alayne & A.T. Grove 1976, Environmental and climatic implications of late Quaternary lake level fluctuations in Africa. *Nature* 261:385-390.
Stuiver, M. & H.A. Polach 1978, Reporting of ^{14}C data. *Radiocarbon* 19:355-363.
Tothill, J.D. 1946, The origin of the Sudan Gezira clay plain. *Sudan Notes and Records* 27:153-183.
UN Development Programme Special Fund 1968, *Semi-detailed soil survey, Gezira-Manaqil areas.* Soil map, Sheet 1. Rome, FAO.
Wendorf, F. & R. Schild 1976, *Prehistory of the Nile valley.* New York, Academic Press.
Wendorf, F., R. Schild, R. Said, C.V. Haynes, A. Gautier & M. Kobusiewicz 1976, The prehistory of the Egyptian Sahara. *Science* 193:103-114.
Whiteman, A.J. 1971, *The geology of the Sudan Republic.* Oxford University Press:290.
Williams, M.A.J. 1966, Age of alluvial clays in the western Gezira, Republic of the Sudan. *Nature* 211:270-271.
Williams, M.A.J. & D. Adamson 1973, The physiography of the central Sudan. *Geogr. J.* 139:498-508.
Williams, M.A.J. & D. Adamson 1974, Late Pleistocene desiccation along the White Nile. *Nature* 248:584-586.
Williams, M.A.J. & D. Adamson 1976, *The origin of the soils between the Blue and White Nile Rivers, Central Sudan, with some agricultural and climatological implications.* Nat. Council for Res., Econ. and Soc. Res. Council Occasional Paper 6. Khartoum: 1-44.
Williams, M.A.J., A.H. Medani, J.A. Talent & R. Mawson 1974, A note on Upper Quaternary mollusca west of Jebel Aulia. *Sudan Notes and Records* 54:168-172.
Williams, M.A.J., J.D. Clark, D.A. Adamson & R. Gillespie 1975, Recent Quaternary research in central Sudan. *Bull. ASEQUA,* Dakar 46:75-86.

DONALD A.ADAMSON
Macquarie University, North Ryde, Australia

10

The integrated Nile

10.1 INTRODUCTION

The Gezira occupies the central position in the Nile basin at the junction of the Blue and White Nile. This chapter places the central Sudan in context with the Nile basin as a whole and attempts to resolve conflicting conclusions which have arisen about the Tertiary and Quaternary history of the Nile. As pointed out in the conclusion of Chapter 9, the evidence left by the Blue and White Nile in the Gezira and adjacent regions must be consistent with the evidence both upstream in the headwaters and downstream in Egypt and the offshore delta.

10.2 EARLY HISTORY OF THE NILE BASIN

In our view the Nile drainage system originated with the rifting of Africa. The Nile rises on the western flank of the Ethiopian Rift and in the inter-arch basins of the East African Rifts. It flows northwards along the downwarped western flank of the Red Sea Rift to the Mediterranean. Tectonic events in the Tertiary associated with the separation of Arabia and the internal splitting of Africa have created the high altitude catchments of the Nile and determined its northward course to the Mediterranean. The directions of rifting, the location and orientation of the major sedimentary basins, and the detailed courses of the rivers in the Nile basin are frequently controlled by geological structures of far greater antiquity than the Tertiary rifting process (Adamson & F. Williams 1980). Medani & Vail (1974) pointed out the relation between ancient geological structures, Tertiary vulcanism and river behaviour in the Sudan, and Mohr (1974) and Black *et al.* (1974) demonstrated the importance of wide-ranging ancient lineaments and faults for the Tertiary rifting events of eastern Africa. Adamson & F.Williams (1980) related geological structures and tectonics throughout the Nile basin to the orientation of streams and sedimentary troughs, and to major changes in the former extent of the basin. It is clear that tectonic events of great age, probably dating back to the Precambrian, have exerted a powerful influence over the development of the Nile basin and its river systems.

The age of an integrated Nile drainage system is however a matter of controversy. For example, one view is that the Ethiopian drainage down the Blue Nile and Takazze-Atbara Rivers or their ancient equivalents only reached Egypt via the Main Nile as recently as the Middle to Late Upper Pleistocene (de Heinzelin 1963, 1968, De Heinzelin & Paepe 1965). According to this view, integrated drainage through the Nile basin is of young age. De Heinzelin (1963) proposed that the Nile basin was formerly broken into a series of separate basins, only the most northerly (the Proto-Nile basin) feeding a river following the present course of the Nile in Egypt and in the far north of the Sudan. The more southerly basins drained either to the Red Sea, the Atlantic Ocean or internally. Water and sediment from the Ethiopian highlands was said to have concentrated in the closed Palaeo-Sudd basin in southern Sudan or overflowed to the Red Sea. The implications of this view of Nile history were summarised clearly by Butzer & Hansen (1968:6): 'Consequently, although the Egyptian and Nubian Nile is an ancient river, the Nile hydrographic system as we know it today is of comparatively recent age. The White Nile and the Blue Nile basins certainly did not merge with the Saharan Nile before the early Pleistocene'. Butzer & Hansen themselves (p.78) adopted a more cautious view of the early to middle Pleistocene Nile: 'the Nile record speaks for an Egyptian river rather than an exotic stream such as we know to-day. Whether or not the Nile received water from south of the Sahara, or experienced summer floods of Ethiopian origin cannot be deduced from the existing record'. Butzer (1980a) reiterated this view and favoured the probability of an input from Ethiopia to the Egyptian Nile since the beginning of the Pleistocene. Wendorf & Schild (1976:233-4) acknowledge that an Ethiopian input to the Nile was older than originally suggested by De Heinzelin and they concur with the conclusions of Butzer & Hansen (1968).

An alternative view (McDougall *et al.* 1975, Williams & Williams 1980) is that drainage from Ethiopia via rivers equivalent to the Blue Nile and the Atbara/Takazze flowed to the Mediterranean via the Egyptian Nile since well back into Tertiary times, although the connection may not have been continuous. The Ethiopian highlands originated from uplift that may have begun during the Cretaceous and vast volcanic outpourings occurred in the late Oligocene and early Miocene about 25 to 19 million years ago. Volcanic activity continued to the present day as shown by the rift volcanoes and the Late Quaternary volcanic ash deposits in the Gezira and along the White Nile (chapter 9 of this volume; Williams & Adamson 1980).

Major incision of the Ethiopian swell must have been well advanced during the Miocene by rivers flowing inland, as well as by rivers draining to the Red Sea Rift and the Indian Ocean. The north-westerly trend of the Atbara and the lower reaches of rivers in the Blue Nile system are structurally controlled (Adamson & F.Williams 1980) by ancient geological features which have probably controlled drainage throughout the existence of these watercourses.

The vast discharge of sediment eroded from the Ethiopian highlands was therefore channelled into the central and northern Sudan. Although the huge sediment trap of the Southern Sudanese Trough is large enough to contain the sediment (Adamson & F.Williams 1980) it is too far south in relation to the trend of drainage, the distances, and the gradients. Besides, it has accumulated sediment for an unknown period of time, far in excess of the Tertiary period, from southern Ethiopia, northern Kenya and Uganda, the Congo divide, Darfur and Jebel Marra, western Kordofan, and the Nuba Mountains – a catchment sufficient to fill this vast sump without the additional sediment from the remote Blue Nile and Atbara/Takazze systems.

In summary, we feel that the Nile has been supplying Ethiopian water and sediment to the Mediterranean since well back into the Tertiary, that is, for from several to many millions of years. In the early stages of the evolution of the present drainage pattern, there were probably several separated basins, but the coincidence between the volume of debris eroded from Ethiopia and the volume of the Nile cone in the Mediterranean, together with micropalaeontological and mineralogical evidence from the sediments in the cone, argue for a very early integration of sub-Saharan drainage with the Egyptian Nile (Williams & Williams 1980). The dearth of clear evidence for ancient Ethiopian sediments in the narrow overdeepened Nile valley in Egypt is probably accounted for by repeated intense scouring since Miocene times. The general evolution of the Nile basin has recently been summarised by Williams & Williams (1980), and Tertiary and Quaternary changes to the margins of the basin and the development of the great Southern Sudanese sedimentary trough have been described by Adamson & F.Williams (1980).

10.3 LATE PLEISTOCENE AND HOLOCENE HISTORY OF THE NILE BASIN

The history of the Nile over the last 20 to 30,000 years is fairly clear thanks to a reasonably complete sedimentary record which can be dated by the radiocarbon method, and to the detailed study of these deposits at key locations throughout the basin: near the headwaters of the White and Blue Nile (Kenya, Uganda and Ethiopia); in central Sudan (the Gezira region); in Egypt and Nubia on the Main Nile; and offshore on the Nile cone.

In this chapter we concentrate on two important time intervals in the late Quaternary which show opposed climatic trends and river behaviour, firstly the most recent period of low sea level, world-wide cooling and glacial advance in the late Pleistocene (c.20,000-13,000 yr BP), and secondly the terminal Pleistocene to mid-Holocene period of rising sea level, glacial retreat and world-wide warming (c.12,000-5,000 yr BP). During each of these two strongly contrasting intervals, the evidence about the behaviour of the Nile should be congruent with the evidence about climate throughout the basin. Further, the evidence from the headwaters to the Mediterranean should be consistent, and be capable of being used to support a coherent story of river behaviour. Table 10.1 summarises some of the vast body of evidence from throughout the Nile basin and draws inferences about river behaviour.

Adamson *et al.* (1980) set out to present a coherent view of the behaviour of the Nile throughout its basin over the last 20,000 to 30,000 years. Having shown how the available evidence was consistent throughout the basin, they

Table 10.1. A summary of environmental, fluvial and depositional evidence, and inference, throughout the Nile basin: A) the late Pleistocene arid cool period between c. 20 000 and c. 12 000 BP, and B) the moist warm period in the first half of the Holocene between c. 12 000 and 5 000 BP.

	Ethiopian highlands (Blue Nile/Atbara Headwaters)	Kenya and Uganda (White Nile Headwaters)	Gezira Region (Atbara River similar to Blue Nile)		Egypt/Nubia (Main Nile)	Mediterranean Cone of Nile
A. The late Pleistocene arid cool period between c. 20000 and c. 12000 BP						
Climate-Vegetation	Cold, drier, minor glaciation, extensive periglacial and frost action. Lower tree line and sparser vegetation.	Drier, and cooler at least on mountains. More extensive grassland, less forest, lower treeline. Less swamps.	Arid.		Hyperarid in Nile valley and to west. ?Some run-off from Red Sea Hills to wadis on east side of Nile. Cooler.	* Sea level fell to c.100 m below present level.
Lake levels	Low.	Low. Lake Victoria lacked outlet for at least part of this time.	Dry in depressions in Wad Mansurab (and at En Nahud).		Dry in playas on west side of Nile. No settlement.	* Calcareous ooze deposited in deep water (>400-500 m).
Nile			Blue Nile	White Nile	Main Nile	
a) Total discharge	Low.	Low.	Low.	Low.	Low.	* Coarse calcareous sand and gravel deposited near shore.
b) Maximum discharge	High but brief duration from reduced summer rains, snow melt. Run-off rapid.	Moderate.	High, but flood duration brief.	Moderate.	High, but duration of flood brief.	* Calcareous sediment deposition c.16 000 BP
c) Minimum discharge	Very low. Reduced storage of water by swamps, soil, vegetation.	Very low.	Extremely low, some cessation possible.	Extremely low, probably cessation of flow.	Extremely low, some cessation of flow probable.	
d) Sediment type	Silt, sand, fine and coarse gravel. Extensive mass movement on hill slopes.	Sand, silt.	Silt, sand, gravel.	Silt, sand.	Mixed silt, sand, gravel. Gravel transported from Ethiopia and Sudan.	

		suspended-load carried to Main Nile.	and bed-loads at or near peak discharge.		discharge.	
f) Channel	Incised.	Braided multiple channels on plains of southern Sudan. Incised/confined in headwaters.	Braided, shifting, multiple channels. Shallow and broad.	Multiple inter-dune channels.	Braided, shifting, multiple channels. Shallow and broad rather than deep.	
g) Flood plain	None.	None.	Rapid aggradation of coarse sediments.	Reworking of sand from Blue Nile and Khor Abu Habl.	Rapid aggradation of mixed sediments.	Deposition of coarse sediment off present shore line.
B. The moist warm period in the first half of the Holocene between c. 12 000 and 5 000 BP						
Climate-Vegetation	Warm, moist, high tree line and stabilisation of hill slope erosion by vegetation.	Moist, warmer. Extension of forest, spread of extensive swamps.	Moist.		Arid to slightly moist.	* Sea level rose to about present level by c.6000 BP. * Sapropel, organic ooze and proto-sapropel deposited in deep water. * Grey mud (presumably Nile silts) deposited in shallow water.
Lake levels	High.	High. Overflow of Lake Victoria (and Lake Turkana episodically) to White Nile.	Contained permanent water c.7000 to 9000 BP.		Contained permanent water for part of period especially c.5000-9000 BP. Terminal Palaeolithic-Neolithic settlements.	
Nile			Blue Nile	White Nile	Main Nile	
a) Total discharge	High.	High.	High.	High.	High.	

Table 10.1. (continued)

b) Maximum discharge	High with extended time of maximum discharge.	Remarkably little difference between maximum and minimum discharges due to buffering by Lake Victoria and swamps.	High.	Little difference high.	High.
c) Minimum discharge	Low.		Low.		Substantial permanent flow.
d) Sediment type	Silt, sand.	Silt, clay, suspended load only. Bed-load mainly reworked sand.	Silt, sand.	Clay, silt.	Silt, sand, gravel locally reworked but not transported over long distances.
e) Sediment amount	High at or near peak discharge, very low at minimum discharge.	Moderate suspended load.	High, suspended load at or near peak discharge.	Moderate to low, suspended load.	High suspended load at or near peak discharge.
f) Channel type	Incised.	Incised in headwaters. Multiple channels still common in swamps, otherwise one main channel with minor subsidiary channel(s).	Sinuous, rather stable, becoming incised, single main channel.	Consolidated into one or two main channels.	Sinuous, rather stable, channels one or few.
g) Flood plain	None.	None.	Deposits mainly silt and clay from overbank flooding, progressively reduced by Blue Nile incision.	Similar to Blue Nile.	Little aggradation, mainly incised.

sought to resolve apparent contradictions between Quaternary history which had been inferred for the Nile in Egypt on the one hand and for the several headwater regions on the other. In resolving these apparent contradictions, the evidence from the Gezira and the lower White and Blue Nile proved crucial because this region is central to the Nile basin and is a repository of geological information in the form of extensive sedimentary deposits.

Extensive and detailed studies of archaeology and Quaternary history of the Nile were carried out in Egypt and Nubia prompted by the construction of the Aswan High Dam and its associated irrigation works downstream. A vast amount of information was obtained on ancient Nile sedimentary deposits especially between the Wadi Halfa and Kom Ombo regions. This work led to interpretations of Nile behaviour which were in apparent conflict with the equally detailed evidence which was being obtained about the Nile headwater regions from a study of numerous Ethiopian and East African lakes. The discrepancy was expressed in general terms by Butzer (1980a) '. . . the Nile record would also seem to pose some contradictions. It is essential that interpretations of the East African evidence relevant to the Nile basin be consonant with the basic Nile flood history documented in Egypt'. The problem is most evident in the time intervals being dealt with in this chapter. For example, Smith (1976) dealing with the period between about 10,000 and 3,000 BC (the terminal Pleistocene to mid-Holocene) comments that in Egypt 'the hydrological budget of the Nile itself now tended to be reduced as a result of climatic changes in East Africa'. Yet this was a period when the climatic changes in East Africa were towards more rain and higher lake levels than either before or since (Livingstone 1980, Gasse *et al.* 1980), climatic changes which are hardly consistent with a reduced hydrological budget of the Nile in Egypt. Problems of reconciling the behaviour of the Egyptian Nile with that of its upper reaches were added to by the 'Nile high-water curve' of De Heinzelin (1968) which reflects the absolute altitudes of former Nile floodplains, not the amount of water in the Nile, as recognized by Butzer (1980a).

During the late Pleistocene interval under consideration in this chapter (c. 20,000 to 13,000 BP), there has also been difficulty in reconciling interpretations about the behaviour of the Nile in Egypt with evidence from the headwaters. As stated by Butzer (1980a), 'The vigorous Darau floods remain difficult to explain in terms of the complex climatic zonation of the Nile Basin in Upper Nubia, Ethiopia, and the White Nile drainage components, on the one hand, and the question of peak discharge versus overall flood volume on the other'. How could vigorous floods, moving heavy bedloads of sand and fine gravel into Egypt from the Sudan (and Ethiopia), be consistent with the evidence from the headwaters which showed that lake levels were low, that the climate was substantially drier and that there must have been a much reduced total discharge by the rivers?

Fairbridge (1962, 1963, 1977), by contrast with others working in Nubia and Egypt inferred a late Quaternary climate and a Nilotic regime that is closely similar to that which we now favour. He interpreted (Fairbridge 1963) the late Pleistocene 20,000 to 10,000 yr period as one of 'almost total aridity

Figure 10.1. The relationships between environmental conditions, runoff and erosion in the headwaters of the Blue Nile and Takazze Rivers, and the fluvial and depositional regimes downslope on the Gezira and in Egypt.
a. For the time interval c.20 000 to 13 000 yr BP during the late Pleistocene.
b. For the time interval c.12 000 to 5 000 yr BP during the terminal Pleistocene and first half of the Holocene.

	Rainfall	Temperature	
a. Late Pleistocene c. 20 000-13 000 BP			
Catchment area, Ethiopian Highlands	lower total, with brief intense seasonal storms	lower, frost and snow common at high altitudes	Environmental inputs
	Reduced vegetation cover. Increased bare soil	Cryoturbation and slope instability severe. Snow melt. Vegetation limits at lower altitude, e.g. tree line	Biological and geomorphological consequences
	Rapid runoff and	Intense erosion	
Upper and middle courses of Blue Nile and Atbara-Takazze Rivers	River flow confined in bedrock channels. Flow extremely seasonal, peak flow high but brief	Sediment load extremely high, high proportion of coarse material	Fluvial and depositional regimes
Lower courses of Blue Nile and Atbara; Main Nile. Gezira Plain, Nile valley in Nubia and Egypt	River flow not confined in bedrock channels. Channels braided, unstable, shifting and shallow	Aggradation on flood plain as peak flow wanes. Coarse sediment deposited in and near channels. Coarse sediment shifted long distances by annual flood	

	Rainfall	Temperature	
b. Terminal Pleistocene to Mid-Holocene c. 12 000-5 000 BP			
Catchment area, Ethiopian Highlands	Higher total, with intense seasonal storms. Longer wet season	Higher, no permanent snow	Environmental inputs
	Complete vegetation cover, swamp expansion. Extensive soil development	Cryoturbation and slope instability minor. Vegetation limits at higher altitude	Biological and geomorphological consequences
	Delayed runoff and	Reduced erosion	
Upper and middle courses of Blue Nile and Atbara-Takazze Rivers	River flow confined in bedrock channels. Flow highly seasonal, peak flow extended	Sediment load fairly high. High proportion of fine material	Fluvial and depositional regimes

Figure 10.1. (continued)

	Rainfall	Temperature	
Lower courses of Blue Nile and Atbara Rivers; Main Nile. Gezira Plain, Nile valley in Nubia and Egypt	River flow not confined in bedrock channels. Channels sinuous, few in number (often one only), stable, becoming deeper	Reduced aggradation on flood plain. Fine sediment mainly deposited by overbank flooding. Coarse sediment mainly locally reworked, not transported long distances	Fluvial and depositonal regimes

in Nubia, and semi-arid conditions extending into Ethiopia and Equatoria. At times Nile flow almost ceased. A transitional phase (oscillatory) occurred in 12,000-10,000 BP . . . The early Holocene (10,000-7,000 BP) . . . was a time when rainfall in the tropics and subtropics was much higher than to-day . . . The Nile flow was at least three times the normal of to-day'.

What is now clear is that the evidence from all the various parts of the Nile basin can, in broad outline at least, be accepted; what requires modification and reconciliation are some of the inferences which have been drawn about the past behaviour of the Nile particularly in Egypt and Nubia.

Adamson *et al.* (1980) and Williams & Adamson (1980a: Figure 12.4) pointed out that relatively dry cold conditions in the headwaters, and low total but sharp high peak discharges by the rivers, are consistent with huge sediment input to the headwaters and downstream aggradation as far as Egypt (Figure 10.1a, Table 10.1). The Nile was highly competent to shift large amounts of coarse sediment during the brief but high seasonal flood peaks, sediment which was then dumped in and near the braided, shifting channels as the flood peak declined. This interpretation of stream behaviour reconciles the evidence throughout the basin during the last late Pleistocene period of major world-wide glaciation. During the terminal Pleistocene to mid-Holocene moister and warmer interval, there was in the headwaters more vegetation, more stable slopes and soils, less erosion, and therefore less sediment particularly of the coarse grades being added to the rivers. Discharge by the rivers was higher but exhibited less extreme seasonal fluctuations; relatively stable deeper channels were established downstream on the Gezira and in Egypt; overall, there was far less sedimentation in the narrow Nile valley in Nubia and Egypt than there had been in the previous more arid period (Figure 10.1b, Table 10.1).

There is no doubt that the detailed evidence from the Gezira region accords closely with that from the headwaters of the rivers which supply its water, the Blue and White Nile. Chapter 9 examines in some detail the evidence from the Gezira (see also Williams & Adamson 1980) which shows clearly that the first half of the Holocene and the final couple of millennia of the Pleistocene were generally times of abundant water supply from the rivers, with intervals that were even moist locally. In addition, the 20,000 to 13,000 BP period in the late Pleistocene was locally arid and water supply from the rivers was greatly

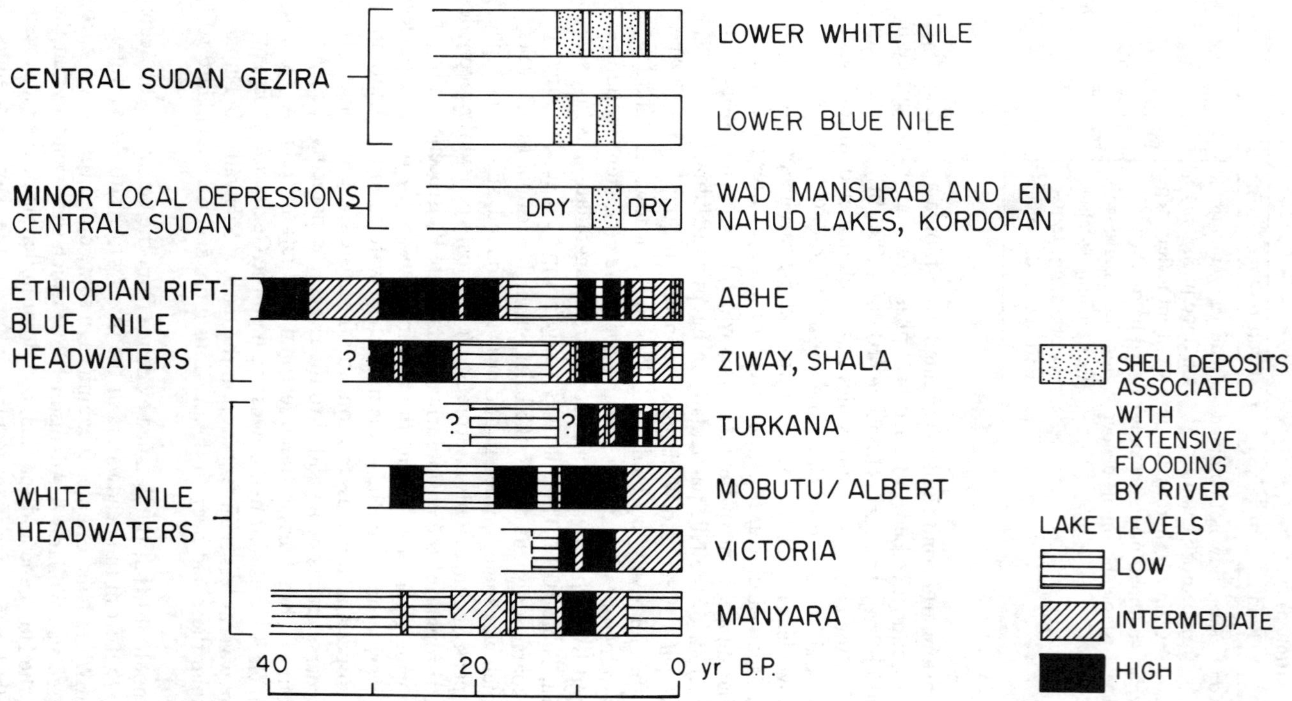

Figure 10.2. Lake levels or intervals favourable for the extensive deposition of fresh-water and amphibious mollusca in the Gezira (associated with the lower courses of the Blue and White Niles), and in the headwaters regions of the two rivers (Adapted from Adamson *et al.* 1980).

reduced. Figure 10.2 summarises much of the evidence from lakes in or near the headwaters of the main Nile branches. Livingstone (1980) sums it up: 'All evidence suggests that the headwater environment of the Nile was relatively dry during the late-Pleistocene, relatively wet during the Holocene' and, as stated by Gasse *et al.* (1980) 'The terminal Pleistocene arid interval now emerges as one of the most distinctive climatic events . . .'

Downstream in Nubia and Egypt it is also clear that the local conditions, at least in so far as overall degree of aridity and temperature are concerned, were generally similar to those further upstream. Figure 10.3 and Table 10.1 summarise a great deal of information from throughout the Nile basin. The behaviour of local lakes and minor ponds west of the Nile in Egypt and northern Sudan (Wendorf *et al.* 1976, 1977, Haynes *et al.* 1979) and in similar latitudes in Arabia (McClure 1976, 1978) clearly show that the late Pleistocene period under review was locally very dry whereas much of the early Holocene was sufficiently moist to maintain permanent water in lakes that were surrounded by abundant prehistoric settlements. Wadis flowing into the Nile from the western desert in Egypt were inactive during most of the late Pleistocene arid interval when conditions were cool enough to allow barley to grow in the seasonally flooded inter-dune hollows where the wadis enter the main Nile valley (Wendorf *et al.* 1979). These wadis apparently became active in the moister early Holocene. This evidence indicates that western Egypt and north western Sudan experienced a cooler more arid late Pleistocene interval and a more humid early Holocene interval, as did the rest of the Nile basin. This evidence is also parallel to that from the playas in the western desert (Haynes *et al.* 1979, Wendorf *et al.* 1976, Wendorf & Schild 1980). However, the situation in Egypt is complex, as wadis flowing from the Red Sea Hills appear to have been active at this time. For example, Wadis Kharit and Shait which enter the Nile valley at Kom Ombo (Figure 10.3) deposited up to 9 m of clayey silt (the Malki Member of the Ineiba Formation) between about 17,000 and 12,000 BP (Butzer & Hansen 1968). This wadi activity may possibly be accounted for by episodic rains localised on the high Red Sea Hills at a time of intensified global circulation during full glacial conditions, a suggestion that is not necessarily in conflict with hyper-aridity and wadi inactivity in the Nubian and Egyptian western desert.

Offshore, the depositional record (Stanley & Maldonado 1977, 1979; Table 10.1) is also consistent with a period of late Pleistocene aridity and reduced total river discharge into the eastern Mediterranean, whereas the sapropels and grey muds of the first half of the Holocene are consistent with an increased freshwater input bearing fine rather than coarse sediment.

Throughout the late Quaternary, the Nile has behaved as an integrated river system with events downstream being determined by those in the headwaters. The characteristics of water discharge and sediment load which originated in the headwaters, in turn controlled the nature of aggradation, erosion and flooding on the low-angle floodplains far downstream. Table 10.1 and Figures 10.1 and 10.3 attempt to collate and summarise the information and reasonable inferences about climate, river behaviour and sedimentation from throughout the Nile basin for both Quaternary time periods considered in this

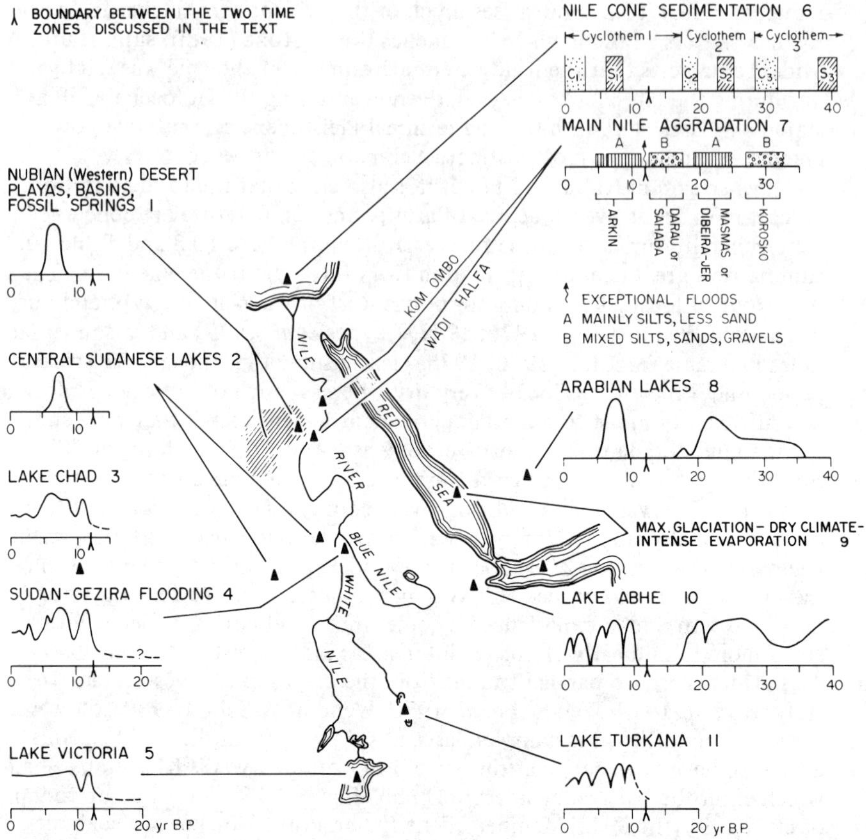

Figure 10.3. Synchronous events in and near the Nile basin indicative of a late Pleistocene arid interval (c.20,000 to 13,000 BP) and a terminal Pleistocene to mid-Holocene moist interval (c.12,000 to 5,000 BP).

1) Wendorf *et al.* 1976, 1977; Haynes *et al.* 1979. 2) Williams *et al.* 1974, Williams & Adamson 1980; Chapter 9. 3) Servant & Servant-Vildary 1980. 4) Adamson *et al.* 1980; Chapter 9. 5) Kendall 1969, Livingstone 1980. 6) Stanley & Maldonado 1977. 7) Butzer & Hansen 1968, Butzer 1980, De Heinzelin 1968, Fairbridge 1962, 1963, 1977. 8) McClure 1976, 1978. 9) Deuser *et al.* 1976. 10) Gasse 1977, Gasse *et al.* 1980. 11) Butzer 1980b.

chapter. These two periods – the late Pleistocene arid period c.20,000 to c. 13,000 BP and the terminal Pleistocene-mid-Holocene moist period c.12,000 to c.5,000 BP – merit particular attention because of the extensive evidence that is available about each throughout the Nile basin and elsewhere in Africa (see Chapter 2), because of their opposite and extreme climates, and because of their close proximity to our own time. Present-day landscapes and available human resources are an immediate legacy of these periods (Chapters 2, 6, 7).

Furthermore, climate in the future will undoubtedly oscillate between the extreme limits which the whole Nile basin has so recently experienced. An understanding of the past behaviour of such a vast river system as the Nile, and of its depositional legacy in the Sudan and Egypt, is therefore of both local and world-wide interest.

10.4 REFERENCES

Adamson, D.A., F.Gasse, F.A.Street & M.A.J.Williams 1980, Late Quaternary history of the Nile. *Nature* 288:50-55.

Adamson, D. & F.Williams 1980, Structural geology, tectonics and the control of drainage in the Nile basin. In: M.A.J.Williams & H.Faure (eds.), *The Sahara and the Nile:* 225-252. Rotterdam, Balkema.

Black, R.M., W.H.Morton & Tsegaye Hailu 1974, Early structures around the Afar triple junction. *Nature* 248:496-7.

Butzer, K.W. 1980a, Pleistocene history of the Nile valley in Egypt and lower Nubia. In: M.A.J.Williams & H.Faure (eds.), *The Sahara and the Nile:*253-280. Rotterdam, Balkema.

Butzer, K.W. 1980b. The Holocene lake plain of North Rudolf, East Africa. *Physical Geog.* 1:42-58.

Butzer, K.W. & C.L.Hansen 1968, *Desert and river in Nubia: geomorphology and prehistoric environments at the Aswan Reservoir.* Madison, University of Wisconsin Press.

De Heinzelin, J. 1963, A tentative palaeogeographic map of Neogene Africa. Viking Fund Publ. in Anthropol. 36:648-53.

De Heinzelin, J. 1968, Geological history of the Nile Valley in Nubia. In: F.Wendorf (ed.), *Prehistory of Nubia* vol.1:19-55. Dallas, Sthn. Meth. Univ. Press.

De Heinzelin, J. & R.Paepe 1965, The geological history of the Nile Valley in Sudanese Nubia: preliminary results. In: F.Wendorf (ed.), *Contributions to the prehistory of Nubia:* 29-56. Dallas, Sthn. Meth. Univ. Press.

Deuser, W.G., E.H.Ross & L.S.Waterman 1976, Glacial and pluvial periods: their relationship revealed by Pleistocene sediments of the Red Sea and Gulf of Aden. *Science* 191:1168-1170.

Fairbridge, R.W. 1962, New radiocarbon dates of Nile sediments. *Nature* 196:108-110.

Fairbridge, R.W. 1963, Nile sedimentation above Wadi Halfa during the last 20,000 years. *Kush* XI:96-107.

Fairbridge, R.W. 1977, Global climate during the 13,500 BP Gothenburg geomagnetic excursion. *Nature* 265:430-431.

Gasse, F. 1977, Evolution of Lake Abhé (Ethiopia and TFAI) from 70,000 BP. *Nature* 265:42-5.

Gasse, F., P.Rognon & F.A.Street 1980, Quaternary history of the Afar and Ethiopian Rift lakes. In: M.A.J.Williams & H.Faure (eds.), *The Sahara and the Nile:*361-400. Rotterdam, Balkema.

Haynes, C.V., P.J.Mehringer & A.Zaghloul 1979, Pluvial lakes of north-western Sudan. *Geog. J.* 145:437-445.

Kendall, R.L. 1969, An ecological history of the Lake Victoria basin. *Ecol. Monogr.* 39: 121-176.

Livingstone, D.A. 1980, Environmental changes in the Nile headwaters. In: M.A.J.Williams & H.Faure (eds.), *The Sahara and the Nile:* 339-359. Rotterdam, Balkema.

McClure, H.A. 1976, Radiocarbon chronology of late Quaternary lakes in the Arabian desert. *Nature* 263:755-756.

McClure, H.A. 1978, Ar Rub' Al Khali. In: Saad S.Al-Sayari & J.G.Zötl (eds.), *Quaternary period in Saudi Arabia*: 252-263. Berlin, Springer-Verlag.

McDougall, I., W.H.Morton & M.A.J.Williams 1975, Age and rates of denudation of Trap Series basalts at Blue Nile gorge, Ethiopia. *Nature* 254:207-9.

Medani, A.H. & J.R.Vail 1974, Post-Cretaceous faulting in Sudan and its relationship to the East African Rift System. *Nature* 248:133-5.

Mohr, P.A. 1974, Mapping of the major structures of the African rift system from ERTS-1 imagery. Smithsonian Astrophysical Observatory Special Report 361.

Servant, M. & S.Servant-Vildary 1980, L'environnement quaternaire du bassin du Chad. In: M.A.J.Williams & H.Faure (eds.), *The Sahara and the Nile:* 133-162. Rotterdam, Balkema.

Smith, P.E.L. 1976, Stone-age man on the Nile. *Sci. Am.* 235:30-38.

Stanley, D.J. & A.Maldonado 1977, Nile cone: Late Quaternary stratigraphy and sediment dispersal. *Nature* 266:129-135.

Wendorf, F. & R.Schild 1976, *Prehistory of the Nile Valley*. New York, Academic Press.

Wendorf, F. & R.Schild 1980, *Prehistory of the Eastern Sahara.* New York, Academic Press.

Wendorf, F., R.Schild, R.Said, C.V.Haynes, A.Gautier & M.Kobusiewicz 1976, The prehistory of the Egyptian Sahara. *Science* 193:103-114.

Wendorf, F. & Members of the Combined Prehistoric Expedition 1977, Late Pleistocene and recent climatic changes in the Egyptian Sahara. *Geog. J.* 143:211-234.

Wendorf, F., R.Schild, N.El Hadidi, A.E.Close, M.Kobusiewicz, H.Wieckowska, B.Issawi & H.Haas 1979, Use of Barley in the Egyptian Late Palaeolithic. *Science* 205:1341-1347.

Williams, M.A.J. & D.Adamson 1980, Late Quaternary depositional history of the Blue and White Nile rivers in central Sudan. In: M.A.J.Williams & H.Faure (eds.), *The Sahara and the Nile:* 281-304. Rotterdam, Balkema.

Williams, M.A.J. & F.M.Williams 1980, Evolution of the Nile basin. In: M.A.J.Williams & H.Faure (eds.), *The Sahara and the Nile:* 207-224. Rotterdam, Balkema.

Williams, M.A.J., A.H.Medani, J.A.Talent & R.Mawson 1974, A note on Upper Quaternary mollusca west of Jebel Aulia. *Sudan Notes and Records* 54:168-72.

Epilogue

The land between two Niles, the Gezira, is the economic and geographic heartland of the Sudan, the meeting place of the two great branches of the world's longest river, and the site of one of the world's largest and most successful irrigation systems. This volume has shown something of the origin and characteristics of this gigantic inland alluvial plain. But what of the future, not only of the Gezira, but also of the rest of the Nile basin? Chapter 10 showed that the Nile basin is an integrated unit with a coherent history from headwaters to the sea. In the future the Nile basin will be integrated in yet another sense as a result of man's intervention. Man's actions in any part of the basin affect the Nile downstream. The most dramatic illustration of this new factor in the Nile basin is the effect of dams and irrigation on the supply of fresh water and sediment to the Mediterranean (Kassas 1972, Rzóska 1976). Not only is the Nile changed, the Mediterranean itself is altered now (Morcos & Messieh 1973, Nielsen 1973) and both the Nile basin and the sea will increasingly be altered in the future by man's works along the Nile. For example, the by-pass canals and associated works in progress in the swamps and plains of the southern Sudan are a present-day intervention of colossal magnitude. Apart from implications for the physical geography of north-eastern Africa and the Mediterranean, man's activities are directed towards the economic and sociological improvement of those who dwell in the countries flanking the Nile. The vast topic of the advantages and the problems associated with these interventions with nature (Fahim 1973, Lemma 1973, Dafalla 1975, Rzóska 1976) is beyond the scope of this book.

REFERENCES

Dafalla, H. 1975, *The Nubian Exodus.* London, C.Hurst & Co., and Uppsala, Scandinavian Institute of African Studies.

Fahim, H.M. 1973, Egyptian Nubia after resettlement. *Current Anthropology* 14:483-485.

Kassas, M. 1972, Ecological consequences of water development projects. In: N.Polunin (ed.), *The Environmental Future:* 215-246. London, Macmillan.

Lemma, A. 1973, Schistosomiasis: the social challenge of controlling a man-made disease. *Impact of Science on Society* 13:133-142.
Morcos, S.A. & S.N.Messieh 1973, Change in the current regime in the Suez Canal after construction of Aswan High Dam. *Nature* 242:38-39.
Nielsen, E. 1973, Coastal erosion in the Nile delta. *Nature & Resources* 11:14-18.
Rzóska, J. 1976, A controversy reviewed. *Nature* 261:444-5.

Glossary

Feddan: unit of area, about 1.03 acres or 0.4 hectares
Gezira: island; hence region of alluvial plains flanked by the Blue and White Nile north of about Kosti and Sennar
Jebel: low hill, or mountain, often rising abruptly from the adjacent plains
Kerrib: gullied margin of floodplain next to large stream or river
Khor: ephemeral or seasonal channel, often large
Maya'a: swampy depression prone to seasonal flooding from major river
Qoz: sand plain, dune or ridge, often wooded
Sudd: extensive perennial swamps of the White Nile and its tributaries extending from about Malakal towards the southern border of the Sudan
Toich: seasonally flooded grasslands bordering the White Nile grazed by Dinka, Shilluk and Nuer cattle during the dry winter months
Zeriba: thorn fence; circular enclosure for domestic animals

Indexes

SUBJECT AND LOCALITY INDEX

AUTHORS CITED